TECHNOLOGY FOR GOOD

# TECHNOLOGY FOR GOOD

## How Nonprofit Leaders Are Using Software and Data to Solve Our Most Pressing Social Problems

JIM FRUCHTERMAN

The MIT Press
Cambridge, Massachusetts
London, England

The MIT Press
Massachusetts Institute of Technology
77 Massachusetts Avenue, Cambridge, MA 02139
mitpress.mit.edu



The MIT Press would like to thank the anonymous peer reviewers who provided comments on drafts of this book. The generous work of academic experts is essential for establishing the authority and quality of our publications. We acknowledge with gratitude the contributions of these otherwise uncredited readers.

This book was set in Adobe Garamond Pro by New Best-set Typesetters Ltd. Printed and bound in the United States of America.

Library of Congress Cataloging-in-Publication Data is available.

ISBN: 978-0-262-05097-5

10 9 8 7 6 5 4 3 2 1

EU Authorised Representative: Easy Access System Europe, Mustamäe tee 50, 10621 Tallinn, Estonia | Email: gpsr.requests@easproject.com

To every person who has ever wondered: What if technology benefited the 90 percent of humanity neglected by for-profit companies because of the immense profits they gain by focusing on the richest 10 percent?

# Contents

# Foreword

Despite the mosquitoes, the unreliable electricity, and the absence of running water or flushing toilets, my summer breaks in rural India remain some of my favorite childhood memories. Far from my home in Illinois, I would settle onto my grandfather's porch to share stories of the technological marvels I had encountered—supercomputers that crunched endless streams of data, handheld Game Boys that opened portals to entire digital worlds, and early cell phones that hinted at a new era of instant connection—any place, any time. My grandfather always listened patiently, asking how these inventions worked, what problems they might solve, and why they inspired such fascination.

But I knew these questions were merely a prelude. Eventually, he would pose the one that mattered most to him—and to me: "How will this technology help someone?" He insisted on asking how it would change the lives of the real teachers, farmers, and workers we had met that very day—people working ceaselessly for a better life despite limited resources, few opportunities, and no safety net.

In retrospect, I see clearly how formative those veranda conversations were. They taught me that technology's greatest promise lies not just in its dazzling capabilities but in its potential to serve and elevate entire communities. All the circuit boards, algorithms, and endless lines of code mean little if they fail to improve the lives of the people my grandfather invoked—people whose daily struggles often go unseen by the commercial tech world.

Today, as I revisit his question, I see both extraordinary promise and a stark imbalance. Many of our brightest minds have access to remarkable tools, from cutting-edge artificial intelligence to sophisticated data analytics. Yet those confronting our most urgent global challenges—climate change, inequality, hunger—often lack access to those tools, capabilities, and structures. The gap between what technology can achieve and what our global community most needs remains as pressing as ever, and my grandfather's insistence on human well-being remains an essential guiding principle.

This is the imbalance that Jim Fruchterman sets out to confront in *Technology for Good.* For over three decades, Jim has shown how powerful it can be to reimagine technology—seeing technology not just as a means of building things but as a way to reshape human experience. His work demonstrates that technology is most transformative when communities themselves design and deploy it to chart their own futures rather than passively receiving the output of distant experts.

I have known Jim through his many contributions to designing a new technological future. At Benetech, he developed software that empowered movements ranging from disability rights to global human rights advocacy. Through Tech Matters, he illustrates how nonprofits and technologists can cocreate scalable platforms aimed at social priorities—from child helplines to climate resilience. In doing so, he underscores the principles woven throughout this book: elevating the voices of the people technology is intended to serve, investing in purpose-driven technical talent, and recognizing technology as an instrument of justice rather than one of extraction.

Yet achieving this vision requires significant structural change. We must fundamentally rethink who has the power to shape our digital world. Today's tech ecosystem often remains driven by a narrow set of commercial interests, with civil society often relegated to the role of end user. This must shift. Social change leaders, nonprofits, and communities need to become coarchitects of their own digital futures, equipped with the tools to deploy technology on behalf of missions that serve the public good.

At the Patrick J. McGovern Foundation, where I serve as president, we have seen firsthand what happens when we connect leading-edge technological capacity with social need. While individual philanthropic initiatives can help, a broad, institutional-level approach to structural support within civil society—creating a robust value chain from technical expertise to equitable infrastructure—can enable nonprofits and grassroots initiatives to operate on equal footing with the private sector.

It is also time to revisit our assumptions about technology's purpose. Rather than evaluating projects by market value or technical sophistication alone, we should measure them by their ability to enhance dignity, expand economic opportunity, and enable political systems to serve all communities more effectively.

In this context, *Technology for Good* offers invaluable guidance. Jim has spent decades in the trenches creating innovations that answer human needs over market demands. His book provides a clear roadmap for organizations ready to explore how technology might amplify, rather than overshadow, social impact.

Jim doesn't confine himself to abstract principles. He presents detailed frameworks for choosing which technologies to embrace, assembling cross-disciplinary teams that marry technical skill with social insight, and creating sustainable models that move beyond pilot projects. By weaving in lessons from real-world wins and setbacks, he helps us see what he sees—adding wisdom to the technical approach and letting us leapfrog years of trial and error.

Jim confronts the difficult questions that every social sector leader must face in this era of rapid technological evolution. How do we ensure that new tools magnify human agency instead of displacing it, especially as AI and automation gain ground? What kinds of governance guard community interests when data have become our most coveted resource? And how do we calculate outcomes when innovation's ripple effects—both promising and problematic—reach beyond the bounds of a standard spreadsheet?

These aren't hypothetical concerns. Their answers will shape the future of social progress in the digital age. We risk allowing technology to heighten inequality, undermine democratic systems, and diminish our collective humanity if we overlook these issues. But I'm also hopeful: if we address them responsibly, we might spark an era of empowerment and shared prosperity. In that sense, *Technology for Good* gives us a framework for action toward a future in which social and technological innovation advance together.

First, we must place human dignity at the heart of every technology project. This requires designing with communities rather than for them.

Their voices, agency, and rights must not only be honored but be actively built into digital systems and solutions.

Second, we should redefine our measures of success. Rather than focusing purely on growth and efficiency, we must examine equity, resilience, and whether people's lives genuinely improve because of our innovations. Meeting this challenge calls for new metrics and methodologies coupled with the willingness to value what cannot always be neatly quantified.

Third, we need a broad societal approach to cultivating a generation of technologists who are propelled by social purpose as much as profit. That requires us to build new bridges between the social sector and the tech world, provide training that ties technical acumen to ethical frameworks, and reinforce the behavior we wish to see—celebrating the technologists who lead and work in the interests of the common good.

Finally, we ought to forge a new digital social contract—one in which the fruits of technological breakthroughs benefit every segment of society, not merely an elite few. Doing this well will call for bold policy ideas, imaginative philanthropy, and collaborations that extend across industries and borders.

These reforms must go beyond idealism, anchoring themselves in the systemic realities where policy, finance, engineering, and civil society all intersect. This book offers practical frameworks that underscore the shared responsibility among philanthropists, nonprofit organizations, governments, and private-sector players. As the examples show, successful initiatives are grounded in methods that can be replicated, sustained, and held accountable for their impacts over time.

Whether you are a nonprofit leader, technologist, academic, or policymaker, the arguments here remind us that technology's direction hinges on the questions we ask and the purposes we pursue. *Technology for Good* invites us to adopt a deliberate, evidence-based approach—one in which ethical design, inclusive governance, and rigorous measurement are not optional but essential. It is precisely by integrating this level of thoughtfulness that we can shape a future defined less by extractive power and more by shared prosperity and genuine human flourishing.

And in that endeavor, there remains a deep well of possibility—proof that when guided by shared values, we can together use technology to anchor an equitable, sustainable, and thriving future.

Vilas S. Dhar
President and Trustee
The Patrick J. McGovern Foundation
http://www.mcgovern.org
March 2025

# Preface

Imagine a Silicon Valley boardroom in the 1980s. Seated around the polished wood table is the board of directors of Calera Recognition Systems, Inc. Each director represents a major venture capital firm with a multi-million-dollar stake in the company, which I cofounded. Today, we are demonstrating a new product to the board for the first time.

Calera was one of Silicon Valley's first artificial intelligence (AI) startups. We trained our algorithms on millions of examples of characters in different fonts to create a machine that could read just about any printed document. Our products replaced the need for people to enter data into computers. Law firms were using our scanner to scan contracts, and insurance companies were using it to scan claim forms.

However, on this day we were showing a social application for the machine: a reader for people who are blind and visually impaired. As a cofounder and Calera's vice president of marketing, I had the honor of doing the product demonstration.

With a flourish, I took a single page of text and fed it into the scanner we had built. It grabbed the page and pulled it through a mechanism that took a digital picture of the page. Four circuit boards with multiple processors and semicustom chips worked their magic, figuring out the letters and words on the page and creating a word processor file. The file was sent to a personal computer (PC) with a first-generation computer voice synthesizer, which read the correctly recognized words aloud.

The voice was far from human sounding, but the words were unmistakable: "Theeze are the timez which try menz souls . . ."

The demo worked!

After a short round of applause, one of our investors asked a question: "Jim, how big is the market for reading machines for the blind?"

I responded: "We think the market is about $1 million a year worldwide." I could feel the energy drain from the room. After an excruciatingly long silence, another investor asked: "And how does this help us get a return on the $25 million we've collectively invested in this firm?!"[1]

My answer that day was not good enough. Our board vetoed the product on the spot. It was a good business decision for investors whom the three cofounders had promised that we'd make them a lot of money—and hadn't yet made good on that promise. But it was a bad decision for society.

It was also a pivotal moment in my career. I wasn't ready to let go of my dream of helping the blind read independently. Months later, I left Calera to start a nonprofit that would create tech for social good, an organization known today as Benetech.

Our first product was the affordable Arkenstone reading system for people who are blind or visually impaired. Soon after we became the leading maker of such reading machines, we expanded into serving people with dyslexia. Next, we created Bookshare, an online library of digital books for people with reading barriers, where the books could easily be made to talk aloud, be displayed with enlarged print, or be converted to braille. Today, Bookshare is the world's largest library of e-books for people with print-related disabilities, delivering more than thirty million accessible books to more than one million people in more than one hundred countries around the world.

For more than thirty years, I have been a leading practitioner of social innovation and cofounder of multiple successful nonprofits focused on using tech for good. That pointed question from my investor on that fateful day has inspired me to ask other questions about what lies beyond profits:

- How can we link the powerful technology invented in Silicon Valley to the needs of people all over the world?
- How can technology be harnessed to maximize good in the face of business considerations that discount good, often to the point of encouraging evil?

- How can we bridge the gap between the profitability of tech and the possibility of making tech that has positive social impact?

At a time when the world desperately needs solutions to large, complex problems that threaten humanity, I wrote this book for social change leaders who are inspired to tackle these big challenges but are unsure how to go about it. My experience shows me that making meaningful progress on any of them will require a comprehensive systems change approach supported by technology—especially software and data. I hope to share what I've learned since that board meeting in Silicon Valley in the hopes that you will be able to do far more with technology for social good.

# Introduction

The accepted wisdom in big business is that the only worthy ideas are ones that make a lot of money, preferably billions. The quest to make money has led to numerous problems in the modern world: the climate crisis threatens all life on the planet; the opioid epidemic in the United States claims more than one hundred lives every day; and declining physical and mental health in lower-income communities around the world jeopardizes hundreds of millions of children.

After a promising start in the previous century, when trailblazing technology inventors and entrepreneurs focused first on making the world a better place and second on making a profit, big tech now is just as obsessed with profits to the exclusion of everything else as any other industry.

I believe there is a better path for technology. For example, what if tech returned to its roots and made people more effective and powerful? Even bolder, what if the benefits of technology also came to the 90 percent of humanity traditionally neglected by for-profit companies in favor of immense profits gained by focusing on the richest 10 percent?

The opportunity before us is to apply the same tools of innovation and company creation used to make billions of dollars in Silicon Valley to improve the lives of billions of people around the world.

Fortunately for those otherwise neglected people, a growing number of innovative technologists want to use their talents for social good. Hundreds of tech-for-good ventures have already been launched to maximize social impact, not profits. Tens of thousands of established nonprofits have figured out that stronger tech and better data can help them accomplish their missions more quickly and more effectively.

What's lacking are enough leaders with the skills to bridge the gap between technology and social innovation. Both are necessary: technology without social innovation will inevitably fail to solve the problem. Social innovation without technology will inevitably fail to scale.

Our future depends on a new wave of tech-savvy social good leaders harnessing the power of technology for positive social change at scale—leaders who see people as the central reason for all tech-for-good projects.

If you care more about using tech to help people and the planet rather than solely about measuring your tech achievements in profits, this book is for you. Whether you lead a nonprofit organization, a government agency, or a purpose-driven for-profit, whether you are a funder, a consultant, a for-profit tech executive, or an individual technologist, understanding how technology intersects with social impact is crucial to your success.

Most of this book deals with how to make a positive social impact with technology. But first, let's start with why.

## THE GLOBAL WHY

To deal with humanity's most challenging problems, the world's governments have agreed to two sets of ambitious objectives deemed crucial to our collective future. First are the Sustainable Development Goals (SDGs) adopted by all United Nations members in 2015. The SDGs are a comprehensive set of objectives covering every kind of social action, including reducing hunger, poverty, and inequality; improving education and health care; increasing access to clean water and affordable energy; and more. We were supposed to meet these goals by 2030, but prepandemic projections pushed this date to 2082, and the pandemic added another decade of delay.

Second is the Paris Agreement, a global treaty around climate change negotiated by the world's governments in 2015. The Paris Agreement calls for keeping global warming to 1.5°C and rapidly decarbonizing the economy. The latest science shows we are already falling far short of making the changes necessary to limit global warming.

Missing target goals on the SDGs and the Paris Agreement will have massive negative consequences for humanity and the planet. It's clear now that business-as-usual approaches will not get us to our shared goals. For-profit companies won't rapidly make such changes because their shareholders and reward systems won't let them. No single government or organization by itself can solve a global social problem. We don't have enough people or money to solve our many problems simultaneously. To have any hope of changing the current trend lines, we need new policies, new programs, new innovations, and new models of cooperation and collaboration among governments, for-profits, and nonprofits for the common good.

In a few words, we need large-scale systems change—change that I believe can be enabled only by using technology properly designed, deployed, and managed to improve the lives of billions of people around the world.

## YOUR WHY

What about you and your organization?

The short answer is that using tech effectively is now a fundamental requirement for leading any organization. If you truly want to help the communities you serve, you owe it to them to do the best job you can, and technology—software and data—is essential to leading a modern team to success at scale.

Ambitious businesses know that success at scale depends on truly understanding their customer: How much does it cost to acquire the typical customer? How much profit are we likely to make from a given customer during the average time they are a customer? How happy are our customers with our product and our service? If our business depends on repeat customers, are they coming back? Are they renewing their subscriptions? These are just a few of the questions businesses use software and data to answer. It would be impossible to run a successful business without knowing the answers—often determined week by week, day by day, or hour by hour. Without answers to these questions, a business might not survive.

Most nonprofits do not have this kind of detailed grasp over their programs. They aren't collecting this kind of data, and the people running the programs usually don't know the details of what is and isn't working.

However, the pressure is increasing on social good leaders to improve this state of affairs. Their staff and volunteers, accustomed to using advanced tech in their personal lives, often despair over the rudimentary tech tools they are using at work. Program managers want to know if their program is succeeding and how to make it better. Donors and social impact investors want to know about the impact their funding is having, and future funding may depend on having strong impact data. Yet most nonprofit organizations and government agencies struggle to access modern tech tools.

Technology by itself can't solve our social problems: we are still a very long way from robots that understand human society. However, we can do amazing things with the technology we do have in terms of making people smarter and more effective. This is a central theme of the book: equipping the people on the front lines of social change with the information and tools they need to make the world a better place.

## WHAT'S IN THE BOOK?

The goal of this book is to deliver actionable advice and compelling stories for leaders who want to create, expand, join, support, and improve social good organizations that see the application of technology as a key element of delivering on their social good mission. The focus is on helping organizations to use technology strategically to advance their programs by strengthening the internal technology powering their operations, to use technology to support their programs, and to build technology solutions as their program deliverables.

In this book, I won't be focusing on equipping your team with basic productivity tech tools, such as email and spreadsheets. I consider such tools to be a prerequisite to doing more exciting tech innovation.

What I will do is share what I've learned as a tech-for-good entrepreneur and the stories of dozens of organizations already using technology for social impact, illustrating my key points with real-world examples. While writing this book, I launched the *Tech Matters Podcast*, in which I interview many great tech-for-good leaders who explain in their own words how they built successful tech-for-good enterprises.[1] In particular, I sought out leaders who overcame specific challenges that cause other nonprofits to struggle with technology. These interviews enriched many of the insights from my own experience that I share in the book.

Of course, tech is a fast-moving target, and today's outstanding tech-for-good organizations may not shine as brightly in a few years. However, I think the lessons they teach have staying power.

In chapter 1, "Bringing Tech to Social Change," I make a case for the need to be tech enabled with the same set of internal productivity tools used by business, government, and nonprofit organizations, including common tools such as email, word processing, and chat solutions. The rest of the chapter and the book focus on applying innovative technology directly to the work of social change organizations. Your field of work defines the kind of technology you need to better operate your programs: community health teams need different software than human rights defenders.

In chapter 2, "The Top Bad Ideas in Tech for Good," I focus on some of the most common ways tech initiatives fail. If you've done any tech in the social change space, you've probably come across these ideas. I maintain that they fail 95 percent of the time. In figuring out why they usually fail and what the 5 percent that succeed have in common, you should acquire a pretty good radar for bad ideas when they are pitched to you!

In chapter 3, "The Three Best Tech-for-Good Ideas," I answer the obvious next question: What are the best ideas for applying tech for good? I focus on three: equipping not just your frontline workers but also your community members with better tools; being willing to radically innovate by killing the dinosaurs in your organization or the field; and embracing cloud computing. These themes will show up again and again in the stories of successful tech-for-good organizations.

Tech-for-good work today benefits from huge advances in design, and chapter 4, "Designing Tech for Good," expands on the need to shift the design of both tech for good and programs to methodologies such as human-centered, agile, and lean design. These terms may have acquired a lot of buzz, but they have the benefit of actually working better than previous approaches!

Chapter 5, "Building Successful AI Solutions," explores the incredibly hot topic of AI and machine learning. By training algorithms to take over repetitive tasks, you should be able to find exciting opportunities for tech innovation. If you focus on making human beings smarter and more effective, you are much more likely to succeed with AI.

I strongly believe that technology efforts cannot have large-scale impact if they are short-lived projects. Chapter 6, "The Tech-for-Good Life Cycle," takes you through the life cycle of the long-lived sustainable tech-for-good effort, whether you are building tech solutions for your own organization or you are creating tech products for other organizations in the social sector.

Chapter 7, "Funding, Talent, and Intellectual Property," describes the three key ingredients common to any successful tech initiative. The high cost of creating tech products can be justified by the low marginal cost of sharing them with hundreds, thousands, or millions of people if you're successful. The right people building the needed products in the right way will probably require millions of dollars of sustained funding if you are based in a wealthy economy. And if donors are spending millions, it probably makes sense to make your technology available under an open license to the rest of the world!

Chapter 8, "The Future of Tech for Social Impact," underscores what is possible with today's fully connected world, where we can take advantage of global connectivity via the cloud to listen to the people we serve, shifting power to them as they do the hard work of social change. We need to ethically use far more data for good than are being used today, and this will drive a large wave of successful AI-for-good applications in the coming years.

The afterword includes my call to you and all the other stakeholders in the social sector to go forth and use technology to build the better world we and our children and our children's children need.

Let's get started!

# 1 BRINGING TECH TO SOCIAL CHANGE

Technology has dramatically changed daily life in modern society. How we communicate, how we get our news, and how we buy and sell products and services have changed over the past fifty years thanks to technological advances. However, poor people around the world do not have the same tech tools as the wealthy. The nonprofit sector is similarly disadvantaged compared to global for-profit business, often a decade or more behind the times when it comes to technology. I call this disparity the social sector time machine! It's a one-way time machine, where the only question is how far in the past the social sector will be stuck.

Visit a nonprofit or a government agency. The odds are quite good that the hardware sitting on the desks is outdated. The software is likely more than ten years old. If there are technical instruments, they are quite likely to be broken as well as obsolete.

The problem is not just that the tech is old. The way programs are designed is also from an earlier era, when experts spent time planning what was going to be done without putting much thought to whether the plans would work in practice or not.

There are several reasons for this sorry state. Let's start with the three main ones. First, the nonprofit sector never has enough money. Being focused on their mission means that better tools for the staff and volunteers of a nonprofit might not be prioritized compared to delivering more services to people in need. Second, the lack of money also makes the nonprofit sector a less attractive market for tech products, so the team may end up using

Microsoft Excel or Google Sheets instead of having a tool that is built to meet the needs of their field.

Third, tech is also not prioritized by donors. Tech tools are often labeled as an overhead expense, which is discouraged by donors. Many donors limit overhead expenses (costs such as accounting, fundraising, senior leadership) to a percentage that rarely covers the actual costs. This policy is, of course, highly regressive, but it is also extremely common. So nonprofit leaders get the message and limit their tech investments rather than argue with their donors about overhead.

That tech is not prioritized is obvious to nonprofit teams. They have access to the latest technology on a personal basis but are asked to use tech tools at work that seem Stone Age by comparison. However, as long as new technology is well socialized and makes their work lives easier, the people who work in social change should be natural allies to tech improvements.

## APPLYING TECH TO SOCIAL CHANGE

The good news is there is a tremendous opportunity to apply technology to social problems. Also, being behind the times benefits those of us in the social sector in several ways. The venture capitalists and tech entrepreneurs spend billions on advancing new technology and products, and most of that money ends up wasted. This is the nature of high-tech investing, but the system works because some portion of these investments lead to the products that succeed commercially. The gains from the few successes offset the losses from the many failures. Thus, when a new technology comes along, it makes a lot of sense for social good leaders to stand back and let the tech industry spend the money to experiment and learn what that new technology is good for. By the time the dust settles, it's likely that some solid products have been created and will be around for a while.

The products that have the broadest appeal generally become the most affordable. This is why most workers in the nonprofit sector have their own powerful smartphones: the tech companies' opportunity to sell hundreds of millions of devices drives prices down and capabilities up.

## THREE DIFFERENT WAYS TO USE TECH

There are three different ways to use tech as an organization dedicated to positive social impact at scale. The first is to adapt tech solutions to power your organization's programs when the common products don't meet your needs out of the box. The second is to go deeper on internal technology by building solutions for your programs. The third is to choose to become a tech company and build products for people in need and/or the organizations that serve them to use.

There is no right or wrong approach among the three, although they almost always involve spending more money as you progress from the first to the last. The question is figuring out which approach is the current best match to the tech problem you want to solve. It is also worth pointing out that tech projects exist on a continuum rather than offering starkly different approaches: there are and will be projects that straddle the lines.

As you examine your options, you will need to keep in mind the people you serve and the people on your team. The whole point of the tech-for-good field is to prioritize people over profits. Because developing technology is expensive, you should be looking for the least expensive way to meet your organizational needs. A heavy investment in tech by a nonprofit organization needs to be justified by a corresponding payoff in terms of cost savings or mission impact or both.

However, before you adopt any of the three approaches, you must first tech-enable your organization by taking advantage of common tech products that are widely available.

## USING COMMON TOOLS WELL

Every social good organization should start off with the basic modern tech tools. Many types of internal productivity tools and solutions work well for every kind of organization, including nonprofits, government agencies, and businesses. These tools tend to be focused on administrative tasks that are much the same in any organization (for example, email) rather than on program tasks that are unique to an organization's field of work (for example, community health).

The existing tech industry does a fine job of meeting common administrative needs, and social good organizations will mostly do fine with what is on offer. Whether it is hardware, software, or services, if millions of people are using the same products, they are likely to be affordable and of good quality. Your first priority when it comes to technology is to equip your team with these basic tools to effectively carry out the everyday tasks common to so many other organizations.

You should not be creating your own technology if there is already a strong set of affordable options that address your needs for email, word processing, accounting, and similar products. You should be buying the hardware and renting the software your team needs to be productive on a day-to-day basis. In tech procurement, these products are described as "off-the-shelf." That iPhone or Android smartphone or laptop was mass-manufactured, and you can simply go to the store and buy it off the shelf—even if the shelf is in some warehouse at an online retailer such as Amazon! The same term applies to many software products, where all you need to do is create an account online and provide a credit card to start using Microsoft Office, Slack, Zoom, or hundreds of other common software applications.

These are examples of "horizontal" tools—basic technology that helps people be more productive regardless of their field. The great majority of organizations use either Google or Microsoft Office for productivity applications and to host their email. There is no need for a human-rights-specific word processor: general-purpose word processors provide that capability for users in every kind of organization, from for-profits to nonprofits to government agencies and more. Tools that make individuals more effective and productive are widely available and relatively affordable because of the huge size of the market. A powerful smartphone is available in much of the world for less than $100, which is why there are billions of smartphones around.

Beyond the tools used by staff on a regular basis, the typical organization also needs a range of software solutions for core business activities. These tools are used by certain staff in functional areas needed across organizations. For example, every organization needs accountants and accounting and payroll software. The typical US-based small nonprofit uses the same accounting software used by the typical small business, QuickBooks, because it can be easily adapted to nonprofit accounting, and most accounting staff already know how to use it. Every organization needs a website, and there are many commercial providers of website design and hosting services, all built on top of standard products such as WordPress.

The vendors of common software products designed for organizations typically make strong efforts to ensure their products are well designed, and they invest in training materials and technical support services to keep their users happily using their products. For popular tools in wide use, you usually can find new employees who already have experience with these software applications (or similar ones). This kind of experience is very helpful: learning a completely new piece of technology can be challenging.

A basic implementation choice for tech is "make or buy." In my view, most nonprofits should be buying their tech, not making it. Of course, "buy" often means "rent." Many modern software solutions (such as Microsoft Office or Slack) are based on cloud platforms that don't require users to buy the product or pay a one-time license fee. Users instead pay each month or quarter or year to use the software, and the vendor operates the software and hosts the users' data. The vendor is paid to ensure the platform keeps operating and stays regularly updated to keep the data secure. In my experience, more and more organizations, both commercial and nonprofit, are choosing to operate their software on cloud platforms because of these platforms' functionality and affordability.

Many of the major for-profit tech vendors have special programs to make their products available at a discounted price for qualified nonprofits. A great resource for these programs is TechSoup Global, a very successful tech-for-good social enterprise. TechSoup has a huge catalog of hardware and software products from a wide array of vendors, such as Microsoft, Adobe, and Zoom. Your organization provides TechSoup with proof of your nonprofit status, and you then qualify for these tech product discounts. TechSoup also offers free and inexpensive resources to nonprofits to help them make the best use of their technology.

Many of these products require some initial setup before use, but then they usually operate without needing much ongoing attention. Because these products are widely available, there are also numerous service providers to assist you with them. Whether to help with installation and setup, training, upgrades, or operating a tech support help desk, vendors are standing by.

The tech needs of nonprofits are not identical to those of for-profits. However, certain tasks are common to a huge number of nonprofits, which creates a big enough market for multiple vendors to develop products to meet those needs. These markets behave much like the market for accounting software: your nonprofit should be able to find a suitable off-the-shelf product.

For example, most nonprofits need to raise money from donors, and a host of vendors offers solutions for fundraising. No nonprofit should be building software for fundraising: it's far better to pick one of these existing products once your list of donors exceeds the capacity of a modest-size spreadsheet! Fundraising software is widely available from both for-profit and even nonprofit suppliers.

In a few large nonprofit fields, such as health care and education, nonprofit organizations or government agencies have greater funding and can justify spending serious money on commercial tech tools specific to their fields. However, with a few exceptions, for-profit companies tend to focus their tech tool efforts for nonprofits based in developed economies, which means that these tools are designed to meet the needs of rich-world organizations and markets and that they are priced for the budgets of these organizations. Thus, commercial options are focused on the top 10 percent of the world's economies, so we still have significant needs for tech solutions that will work for the other 90 percent. For example, social workers based in the United States have many commercial case management tools to choose from, but those tools are not a good fit for the rest of the world because they are often too expensive and too customized to the unique requirements of the US market.

Once you are using common tech tools well, you can consider which of the exciting approaches to using tech to support your programs works best for you.

### OPTION 1: ADAPTING TECH SOLUTIONS TO YOUR PROGRAMS

Ninety-nine percent (or more) of nonprofits should be looking for effective tech partnerships to get the technology specific to their needs. They shouldn't be doing extensive software development any more than dentists or restaurant owners should be. Unfortunately, they might have to do some or put up with less capable solutions because of market failure in the development of most social good apps.

The most straightforward way to getting the tech you need is to adapt a standard platform. The platform itself should have many of the capabilities you need, but you may have to bridge any gap for your particular requirements.

A common approach is to outsource the adaptation project to a for-profit tech consultancy. For common platforms such as Salesforce, a large number of such consultancies are standing at the ready. Another approach is to find a nonprofit tech partner that has already built a solution for nonprofits like yours and will help adapt that solution to your needs.

#### Customizing a Platform

Doing modest customization to an existing tech product or platform is often a cost-effective way for nonprofits that need a solution to meet their specific program needs. Many tech platforms require some level of setup for each specific organizational customer. You will probably already be familiar with this requirement from your organization's use of horizontal tools such as accounting and human resources software, where the software is adapted for your chart of accounts or your recruiting approach. Customizing a platform is similar but more involved.

This customization approach brings many of the benefits of using a standardized product as long as the amount of customized work is small compared to the capabilities of the base platform. The product remains mostly the same as that used by many other organizations, and it is maintained and kept secure by the platform vendor. You are responsible only for maintaining the custom bits created for your particular program or organizational need.

For example, many nonprofits end up with customized solutions built on top of common business platforms such as those from Salesforce, Microsoft, and many other vendors. This is a very common approach to needs for customer relationship management (CRM), even if many nonprofits do not think of the communities they serve as "customers."

I faced a CRM challenge when I was running Benetech, my first tech nonprofit. We had created the online library Bookshare to meet the needs of people with disabilities (more on Bookshare later). Our number of users kept growing, and our tech support team was having a hard time keeping up. My engineering team wanted to build a custom tech support platform for our readers with disabilities and special education teachers, but that would have taken time away from working on new software features specifically to help our target users. We instead adapted (with help from an outsourced firm) a commercial tech support platform based on Salesforce, which saved us a great deal of time, money, and distraction, while giving us a decent tech support solution for our team and our users.

Just because you can make the software internally doesn't make doing so a good idea. In Benetech's case, it made more sense to pay to use a commercial off-the-shelf product such as Salesforce, even though it was far from cheap to do so, rather than spending an inordinate amount of time and money creating our own from scratch.

When outsourcing tech work, it is very helpful to have someone on your team who is experienced with managing tech projects. An experienced project leader doesn't have to be deeply technical but must understand enough about how to set up such an effort for success. Such leaders are also aware that most tech customizations are going to need ongoing maintenance and tweaks and so will build that need into the planning and budget for a tech project.

### Nonprofit Tech Partners

Finding a nonprofit tech partner that has built a solution for nonprofits like yours works well when the field in which you work is big enough. Over the past twenty years, many tech-for-good nonprofits have launched to solve a common problem in a field by creating a product for nontechnical nonprofits. One of my goals for this book is to encourage the creation of far more tech-providing nonprofits, and I will share more of their stories.

The dynamic between a nonprofit organization that needs tech to enable its work and a nonprofit that makes tech for that field can often be easier than a relationship with a traditional, commercial tech vendor. The incentives are often better aligned when two groups with compatible missions partner together. A nonprofit tech vendor is focused not on profits but on creating products that have the most social impact. These tech nonprofits frequently decide to release their software as an open-source product, which means the software is available for reuse and improvement to anyone without a license fee.

Global public health is an example of a field with numerous nonprofit tech options designed to meet the needs of public health organizations. National health ministries in eighty countries use the open-source software DHIS2 (formerly District Health Information Software), created by a team at the University of Oslo, as the core software for operating their clinic networks. The public health nonprofit VillageReach created software for managing the logistics of medical supply chains, which it turned into another open-source project, OpenLMIS (Open Logistics Management Information System). Basically, this software replaces paper-based systems for managing warehouses and transportation to make sure that vaccines, medicines, and other critical supplies are available in the clinics where they are needed and when they are needed. In subsequent chapters, I share the stories of even more great tech-for-good nonprofits operating in the global public health field.

Beyond public health, many other fields have their own nonprofit software vendors. In climate change, for example, MapBiomas (which I discuss more later) has technology solutions to make it easier for countries to have their own AI-powered communities working with satellite data to understand the evolution of local land use. The nonprofit I currently lead, Tech Matters, created the open-source platform Aselo, which already helps more than fifteen national child helplines with modern contact center software, which I expand on in the next chapter.

Thus, before embarking on building your own tech tools for your nonprofit, it is well worth checking for a nonprofit tech partner that is already working on something similar.

### OPTION 2: DEVELOPING YOUR OWN TECHNOLOGY SOLUTIONS

What happens if using or adapting an existing solution is not an option for your organization? Or when there simply isn't an existing tech solution that closely matches up with a critical need for your mission? What if your team is simply miserable trying to make the existing for-profit tech product work for their needs?

Although building your own tech solutions is a terrible idea for a small community-based nonprofit, it is increasingly common among larger ambitious nonprofit organizations that can find the funding for tech development. If you aspire to change the lives of millions of people for the better, there almost certainly will be strategic use of such tech. Where the social need for better technology is urgent, and the existing technology falls short, you need to take the extra step of creating new technology. Developing your own technology means that your organization is partly or wholly a tech company, albeit one probably organized as a nonprofit. Becoming a tech-developing organization is not a decision to be undertaken lightly: it involves a considerable commitment of resources as well as integrating technologists with specialized skills into your organization. Also, tech has a different type of culture than that of typical community-based nonprofit organizations, which can create tension inside a nonprofit organization when these two different cultures are both present.

Tech is expensive to develop. You are building something that needs tech people, who are very costly by nonprofit salary standards even if your tech staff agrees to take a significant pay cut to join your nonprofit to do more meaningful work. Immense amounts of profit potential enable tech companies to pay individual developers hundreds of thousands of dollars in annual salaries. Thus, custom tech for nonprofits tends to be concentrated at tech-developing nonprofits that are building products for hundreds or thousands of other organizations, at newer nonprofits where the tech is central to their program design, or at an established organization that can justify the expense to create or upgrade the technology of a large program.

### Trust Is the Currency

Central to any technology development are the questions of who the potential users are and what their unmet needs are. But there is an important twist to these questions when you are developing a product in the nonprofit sector rather than the for-profit sector.

I believe trust, not money, is the currency of the social sector. Your users need to trust your intentions to serve them. They are entrusting you with their data, the data of vulnerable people, with the hopes and expectations that something worthwhile will happen as a result of your use of those data. You and your team will need to earn that trust over time by listening to people who might feel that nobody listens to them and then honor that trust by building tech products that signal that you listened. I explore designing tech-for-good products in much more depth in chapter 4, "Designing Tech for Good."

The most common kind of technology development in the social good sector is program technology. Program technology is created specifically for a single organization's programs and is not meant to be shared with other nonprofit organizations. It might be a tool designed just for the internal staff's use or a tool used to deliver services to the community served by this specific nonprofit. In the latter case, you are usually designing for two different groups of users at the same time: your internal team users and your external community users. Organizations should ideally have internal capacity to create these solutions, which means they need to bring skilled personnel onto the team, such as product managers, user experience designers, and software developers (if the needed tool is a software application).

Many nonprofits embark upon a major tech-development adventure by outsourcing 100 percent of the tech development to a for-profit vendor. This approach often does not end well. I frequently run into nonprofit leaders who are being held hostage by the vendor they paid to develop the software they need to use to run their programs. The vendor negotiated the contract so that they own the rights to the software, not the nonprofit that paid for it. In these circumstances, the software ends up costing much more than originally anticipated, and every adjustment requires more money, assuming the vendor can be bothered to make the change.

At the same time, I am also familiar with many stories of nonprofits who were very well served by their for-profit vendor partners. Many of these for-profits chose to partner on a pro bono (free) or a "low bono" (deeply discounted) basis, as is often the case for larger tech companies that are part of the Pledge 1% movement, in which the company's founders dedicate company stock, product, and volunteer time to help nonprofits. In all cases, hope for the best but plan to protect your nonprofit mission whenever you work with an outside for-profit company for your technology.

Even if your organization doesn't want to build an internal tech team (or can't afford one), I strongly recommend you have at least one strong tech person whose job it is to represent the interests of your organization and minimize the chances of being exploited. Your own tech expert on the team will greatly increase the chance of finding a vendor who will be a good partner and proudly support your mission. You should also negotiate a contract that protects your organization from being held hostage by a vendor.

Organizations that create program-specific technology are more numerous than many people in the social sector realize. This is not surprising, given that almost all major businesses now develop large portions of their own technology. If you peel back the covers on a bank or an insurance company or a major retailer, you will find a bunch of tech people writing software and analyzing data. Some of the most famous for-profit tech startups in recent years, such as Uber and Airbnb, are in this camp. Even though they are on the outside a ground transportation company and a short-term housing rental firm, respectively, on the inside, these two organizations are software companies that have used their software and data prowess to shake up an unsuspecting industry.

The same thing is true of many of the new and exciting nonprofits, such as more than 60 percent of those winning the Skoll Award for Social Innovation recently (2018–2023).[1] Founded by eBay's first president, Jeff Skoll, the

Skoll Award is widely regarded as the top award in social entrepreneurship and currently comes with $2 million of unrestricted funding. Even though many nonprofits may not position themselves as tech companies, they need technology to power their innovative programs. Here are some of the recent Skoll awardees that have developed their own technology:

- ARMMAN (providing prenatal support in India)
- Callisto (supporting survivors of sexual assaults on US university campuses)
- Center for Tech and Civic Life (supporting local voting administrators in the United States)
- Code for America (helping make social services more accessible in the United States)
- Crisis Text Line (delivering peer mental health counseling via text, founded in the United States)
- MapBiomas (tracking land use change, founded in Brazil)
- MyAgro (supplying layaway finance for smallholder farmers in Africa)
- Noora Health (supporting family caregiving in Asia)
- Organized Crime and Corruption Reporting Project (supporting journalists who investigate international organized crime and corruption, based in Sarajevo)
- Reach Digital Health (focusing on maternal health, founded in Africa and formerly known as Praekelt.org)
- Thorn (combating online child sexual abuse)

The common thread to these organizations is that their social innovation is closely tied to their ability to use technology to power that innovation. They simply could not have social impact at scale without the internal capacity to build and operate the needed technology. That mission-based urgency justifies the considerable effort to create new tech.

### OPTION 3: CREATING A NONPROFIT TECH COMPANY

The last group of tech-for-good organizations make technology for other people and organizations to use. I mentioned them earlier when I suggested that nontechnical nonprofits should look for nonprofit tech vendors in their field. They look like tech companies but are usually organized as charities. They produce products that they sell or give away and generally treat their users as both customers and partners in social change. The products can take all sorts of forms: software solutions, hardware devices, data products, and services. Even more so than the organizations that develop tech for internal program use, these tech-providing organizations have to build their own teams of technologists. After all, they are tech companies that focus on social good. Instead of trying to make maximum profits, like the typical for-profit startup, they want to make the maximum social impact possible.

These nonprofit tech companies exist to remedy widespread market failure. They are created when a critical mass of people or organizations needs the same thing but the market is just too small to tempt venture capitalists to put money into that product. This is quite common in social good fields where the tech needs are different from those in other fields. These field niches are typically known as *verticals* in tech business jargon. Examples of business verticals include restaurants, dental offices, golf courses, and banks. In social good, we might call these groups "fields" or "communities of practice." There is something similar enough about the social programs provided in each of these fields to make vertical products feasible.

Just like running a restaurant is different than being a dentist, being a human rights worker is different than being an agricultural extension worker. The for-profit tech industry does a good job of meeting the needs of restaurateurs and dentists in a wealthy economy such as the United States but not a particularly great job of meeting the needs of human rights defenders or rural agricultural advisers throughout the world. As already noted, the for-profit tech industry avoids markets deemed insufficiently lucrative, whether organized by field or by geography.

As a technologist, I find meeting the needs of social good verticals particularly rewarding when the work is fundamentally about helping communities and the horizontal solutions in common business use are not the right answer. The right technology can unlock a world of human potential!

**The Arkenstone Reader: Learning on the Fly!**

I started a nonprofit tech company very much by accident. My social good career launched soon after that fateful board meeting with my venture capital investors, described in the preface.

Our for-profit tech company, Calera, had created what was then a pioneering AI product that did optical character recognition that worked well. It could read the text off a printed page and turn it into a digital file. This AI technology had strong commercial applications in business and government because it replaced the need for people to do the boring tasks of keying data into forms or retyping documents.

However, the socially worthwhile application of optical character recognition was to help the blind to read. It had been a dream of mine since my college days at Caltech, ironically, when I first learned about the application of optical pattern recognition to guiding smart bombs. My college brainstorm was to come up with more socially beneficial applications of this cool technology than the making of smarter bombs. My idea was that recognizing letters and words instead of tanks on the battlefield would be a great way to use pattern recognition. From the time I cofounded Calera, my first venture-backed startup, I had dreamed of using our technology to help the blind. This led to the development of the reading machine prototype as a secret project of my marketing team and the engineering team led by my cofounder Dave Ross (my former boss from the first private US company to get a rocket to the launch pad, but that's a different story).

At the time we built our prototype, there was only one reading machine on the market, from Xerox, and we thought our new reading machine for the blind would be better and much more affordable. However, the fact that Xerox was making only around $1 million a year selling their Kurzweil reading machine was the fatal flaw in our plans: Calera's board vetoed the project in that same meeting where it was first demonstrated because of the tiny market. Other than as a social good project for a giant company such as Xerox, a reading machine for people who are visually impaired just didn't seem compatible with a for-profit company.

Following that board meeting, I went to our tech law attorney, Gerry Davis. Gerry was a fan of helping the blind, and when he heard about the board setback, he had a new path in mind for my project. "I can help you start a deliberately nonprofit tech company, Jim!" he exclaimed. I laughed because Calera, like so many other for-profit startups, was accidentally "nonprofit" (i.e., it was still losing a lot of money).

A few months later, I quit Calera after eight years there as a founder. Our venture capitalists were worried I would hire away Calera's engineers and compete directly with it via a new for-profit startup. I instead offered not to compete with them and not to hire away their engineers for a year if they would agree to sell me their character recognition product at a deep discount (and on credit) so I could start selling reading machines as a charity. Of course, at the time I had no idea how to do that!

I took Gerry up on his offer. He incorporated Arkenstone as a tech company organized as a nonprofit, and he did the pro bono legal work to convince the US Internal Revenue Service to recognize us as a charity. Arkenstone was later renamed Benetech, which continues to help people with disabilities to this day.

During Arkenstone's first year, we went out and talked to blind people all over the United States as well as in Europe. My initial idea of having a support network of local volunteer engineers help install the product in their home or workplace didn't go over well with our potential users. They explained that there were many talented blind people who couldn't get hired but were well suited to be independent tech dealers. That model made sense to me: the commercial PC industry had shifted to a dealer model a few years earlier. So we quickly assembled a group of local dealers, the majority of whom were people with disabilities.

Running their own businesses, our dealers sold the heck out of the Arkenstone Reader. The reader turned a standard PC into a reading machine for less than $5,000. That price point was critical. The existing Kurzweil reading machine from Xerox had recently dropped in price from $40,000 into the $15,000–$20,000 range. It was well publicized and many blind people wanted one, but few could afford it.

The less than $5,000 price put our reading machine into the range of a family or an employer buying equipment for a blind employee. Rather than the Arkenstone Reader being in a $1 million per year market, within three years we were making $5 million a year in sales. Surprisingly, our growth was financed by our tech company suppliers: Calera, which made the character recognition PC card; Hewlett Packard, which made the scanner; and Digital Equipment Corporation, which made the artificial voice synthesizer that read the words aloud. By requiring our dealers to pay for the product up front and getting sixty days to pay our suppliers, we were able to generate the cash flow needed to pay our small staff. We were even turning a (very) small profit, which we of course reinvested in improving the product.

Arkenstone was the only tech startup I was ever associated with that surpassed its business plan objectives. Of course, that was because our revenue goals were so modest! Our venture capitalists would consider $5 million a year just as much of a failure as $1 million a year because venture-backed companies need to make much more money than that to justify the investment risk. However, in the nonprofit world a slightly profitable $5 million a year turns out to be a barn burner.

As the years went by, PC technology kept getting better and cheaper. As the cost kept falling, more and more people with a visual impairment could afford to buy a reading machine. So although our revenues stayed around $5 million a year, we were selling more and more reading machines a year over the next ten years as the price per unit fell from $5,000 to $4,000 to $3,000 to $2,000 to $1,500. We were able to replace the Calera hardware with a piece of software that ran on the PC rather than on a co-processor card. As a result, we introduced Open Book, the first piece of talking software for blind people that ran on the then new Windows operating system.

Five years in, we did a big survey of our users and were surprised to find out that 15 percent of them could see fine but had dyslexia and struggled to read. Our product had one feature that the dyslexic users loved: a karaoke-style visual highlighting of the word being spoken aloud by the voice synthesizer (like the bouncing ball in a song video). However, everything else about the product was right for blind people (especially that it was keyboard driven) and wrong for people who are dyslexic (who prefer to use a mouse to click on things). So we created a new reading product designed for users who are dyslexic. Our revenue was enough to allow us to reinvest in improving the product, but we struggled to find the money to do new things not closely related to reading.

After ten years, close to fifty thousand people were using our systems to read. They didn't need to ask for a volunteer or pay someone to read to them. They could simply read the books they wanted to read by scanning them. This was a big shift in the system of making books accessible, where people with print-related disabilities such as blindness, visual impairment, and dyslexia could independently read the books they chose rather than depending on sighted people to read books for them.

### More Nonprofit Tech Companies

Nonprofit tech companies make strong partners of organizations that just want to be tech enabled, and they can also help organizations developing technology for internal program use. Many of the detailed stories in this book are about real-world nonprofit tech companies. Here are just three notable examples, which I expand on in subsequent chapters:

**Kobo Toolbox.** Two Harvard public health professors created a survey tool that works well even where there is no internet. Their tool is now used by more than fourteen thousand different organizations for conducting hundreds of millions of surveys each year.[2]

**MapBiomas.** A Brazilian team used free satellite images and AI technology to analyze land use patterns over decades and subsequently created technology to quickly spot illegal logging activities. They started in the Amazon region and branched out to more than a dozen countries around the world.

**Nexleaf Analytics.** Nexleaf makes network-connected temperature sensors for the vaccine supply chain, most notably for the refrigerators where the vaccines are held in health facilities. It combines hardware devices and software solutions to increase the effectiveness of immunization systems.

From my many conversations with nonprofit leaders, it is clear there are many more needs in the social sector for nonprofit tech vendors to meet.

One of this book's goals is to spark the creation of more such nonprofit tech organizations that can take on some of these common unmet challenges.

## CONCLUSION

The three different approaches to tech for social good—adapting technology for your programs, building your own internal technology, and becoming a tech company that delivers tech products to others—are not rigid or exclusive. You do not need to use just one of these approaches. Organizations can even take on all three.

Adapting and building technology to meet internal needs are the most common ways nonprofits strategically apply technology. Nonprofit tech companies that create social impact tech products is a smaller group of nonprofits, but of course their products are intended to benefit many other social good organizations. Going forward in this book, I share many stories and ideas to help those of you using one or more of these three options for applying tech for social good.

There is no single right path for applying technology for social impact. From the hundreds of nonprofits actively building technology solutions, it is clear that technology can be a breakthrough tool in delivering positive impact in just about every area of social change work.

# 2 THE TOP BAD IDEAS IN TECH FOR GOOD

The first ideas social good leaders come up with for applying tech for good are usually terrible—an outcome they share with leaders of tech for-profit companies, whose first ideas aren't that great either. To be honest, in the case of social good leaders the idea usually isn't their own but one recommended to them by a board member, an adviser, or a tech vendor, and most nonprofit leaders are not experts in evaluating tech ideas.

However, the tech business community is particularly good at winnowing out bad ideas before they're implemented. Investors and entrepreneurs want new companies to create significant value quickly using the fewest resources possible. This requires applying modern design approaches to creating new products, a process that focuses companies on rapidly learning what works and what doesn't. Learning requires getting customer feedback early on and quickly, which makes it far less likely that companies end up with a new product that nobody buys or uses.

The nonprofit sector has been slow to adopt modern design practices, so many bad tech ideas get implemented, and a tremendous amount of time and money gets wasted in this sector. Just because someone thinks an idea for a tech solution should work and that people should use it doesn't mean it will work or that people will use it. Even having a clearly defined important social objective does not guarantee that the tech being deployed will meet the objective. Of course, the people involved don't think the idea is a bad one, which is why they continue to pursue it. They learn that the hard way: through failure.

In my experience, while there are an unlimited number of bad ideas for tech for good, they generally fall into one of four types. Just about every week, I find myself talking a top social good leader out of one or more of these bad ideas. The issue is not that these ideas are *always* wrong; it's that they're wrong at least 95 percent of the time.

### BAD IDEA 1: THE APP NOBODY WILL DOWNLOAD

Mobile has taken over the world. The Pew Research Center reports that 90 percent of humans on the planet today have a smartphone and that another 7 percent have a cellphone that is not a smartphone.[1] Therefore, a nonprofit leader's thinking goes, if a tech nonprofit is going to deploy social good technology that is designed for typical people, it should be a mobile app! Right?

Wrong. Even when people have a mobile phone, it's extraordinarily difficult to get them to download and regularly use apps. This truth was learned years ago the hard way by the for-profit world when it went through an infatuation phase with apps. Thousands and thousands of apps were developed, yet very few of them made money. As a result, for-profit companies today are rightfully cautious about building a new app outside of the few places where such apps have been proven to succeed.

Unfortunately, most nonprofits haven't gotten the memo. Their leaders—or their board members, expert consultants, or donors—still think that apps are a good idea. They spend the money building an app and are crushed when it is seldom downloaded and even more rarely used. Data from the mobile intelligence company Quettra show that commercial apps lose 90 percent of their daily users after just one month on average.[2]

Why is it so difficult to build a successful app? Inertia is a powerful force. People will keep doing things the same way unless a new solution is much better than what they are currently using or they are required to use it. Another challenge with apps is that they need users to take multiple, novel actions to be used regularly. It is hard to justify the effort to download and learn an app for a task done once or only on occasion, especially when the alternatives are going to a known and widely used website, searching the internet, or simply asking a smartphone for an answer. Users are doing those three things all the time and will stick with the familiar.

Yet there are numerous examples of wildly successful apps in many areas. Of course, this list changes every year!

- Social media (Facebook, Instagram, TikTok, X, formerly Twitter)
- Work productivity (Slack, Zoom)
- Messaging (WhatsApp, Facebook Messenger)
- Entertainment (Netflix, Spotify, Apple Music)
- Games (sudoku, *Clash of Clans*, *Roblox*, the latest car-racing game)
- Banking (Venmo, PayPal, your bank)
- News (BBC, *New York Times*, *Washington Post*)
- Camera/photography (built-in apps, Lightroom Mobile)
- Lifestyle (eHarmony for dating, Calm for meditation, Noom for weight)
- Language learning (Duolingo)

These successes share a few important traits: they all offer something a typical user is likely to need, want, or require on a regular basis; they generally are easy to use; many are fun.

An additional factor—not to be underestimated by any nonprofit thinking that an app is the way to go—is that the organizations that develop these successful apps typically spend tens of millions of dollars annually in improving them, adding functionality, and testing them with users. They also promote the apps, so more people know about them. And they use every psychological trick in the book, supported by technology, to hook users, keep them coming back, and increase the amount of time they spend on the app, with success measured in increased minutes of use per day.

### Failed Apps for Social Good

Given the challenges of creating successful apps for social good, there are numerous examples of failures. The reasons they fail are usually not specific to apps but common to the failure of any tech innovation, such as not solving a real problem or serving a real need for real users out in the real world. Here are a few notable failures:

**"Witnessing apps" that aren't around to witness.** Capturing human rights violations or police brutality on camera is critical for justice or reform. However, the capture is rarely made with a specialized app, such as the

American Civil Liberties Union's Mobile Justice app or the Witness.org/Guardian Project CameraV app. Instead, witnesses (and sometimes victims) whip out their smartphones and begin recording video or taking pictures with their built-in camera because the camera is already in their hands, and using it is a common behavior of smartphone users (they can even easily access the camera without putting in their security code). If they are not using a built-in smartphone camera directly, they are probably using a camera-aware social media app such as Facebook or Instagram that runs on their phone. Even though the specialized witnessing apps are much better for the task of recording violence, this advantage does not offset their one huge disadvantage: the people witnessing the violence rarely have such a specialized app on their phone (and if they do, they usually don't have the time to find it and open it).

**Social apps without a community.** Many, many nonprofits want to connect people by building a new social platform without the defects, real or perceived, of mainstream social media platforms. I call this group "Facebook but better." Here are just a few reasons why imagining a better social network is far easier than creating one!

- *The network effect.* Networks with a huge number of users have an outsize advantage over small networks. Facebook, email, and the phone system have user numbers in the billions. When I want to communicate with someone on these networks, I generally can. Imagine inventing something better than Facebook but with just one thousand users.
- *Overwhelming advantage in understanding (and changing) user behavior.* Companies such as Facebook, Google, Apple, Amazon, and so many others have years of experience in tracking their users and mountains of data to mine! They constantly run experiments at huge scale to learn how to maximize whatever objectives they have, which are generally about users using the app more and seeing more ads, which make the tech company more profits.
- *Huge money imbalance.* Having billions of dollars at their disposal means the big tech companies have the ability to build strong

products, innovate quickly, and make life difficult for potential competitors by either buying them or burying them or both.

Tech Matters, my tech nonprofit, has often been invited to bid on contracts to build a new social network for good, a favorite approach of government and United Nations agencies. What happens in the social good sector, though, is that these networks get created but rarely gain traction. The same problem goes for networking websites. People don't have a strong reason to come back when there's nothing much going on compared to the activity of a massive existing social network.

The best versions of social networks for good have more modest objectives than being "better than Facebook." I am on a few low-traffic email or WhatsApp lists of people who have similar narrow common interests—for example, leaders of open-source foundations. The benefits of using something as old-fashioned as email are that nobody has to build a new app and the barrier to participation is low. Further, if the group is talking about something one considers uninteresting, it's easy to ignore.

**Information apps that don't inform.** Modern smartphone users use the built-in search capabilities of their phones and a personal assistant such as Siri or Google Assistant to get information. Smartphones are designed to provide easy interfaces to finding out facts. Specialized apps don't stand a chance against the ingrained behavior of a typical smartphone user. A host of highly successful projects deliver information without needing a new app. For example, the COVID-19 health-alert firm Praekelt.org (now renamed Reach Digital Health) was a WhatsApp-based chatbot that delivered accurate data and answers to key questions in many of South Africa's languages. Because it used WhatsApp, it reached 3.5 million people inside of ten days: a far better fit for the urgent task of pandemic-related information than a stand-alone app would have been.[3]

**Safety apps that aren't easily available to keep people safe.** A safety app is one designed to assist the user when the user is feeling threatened with harm. It does no good if people haven't downloaded the app or don't remember how to use it in an emergency. Even people who are under severe threat generally stop using these types of apps. It's not easy to design an app for situations that occur rarely.

A well-publicized example of a safety app that never achieved critical mass adoption is the Amnesty International Panic Button app. The goal was to alert a network of supporters when a human rights defender was detained by security forces. After five years of intense work with no results, the Amnesty team decided to shut down the app. When they did, though, they also took a step that was incredibly responsible and helpful to other organizations that might consider developing a similar app: they wrote two blog posts on why they were shutting down the app.[4] The debrief focused on two big problems: false alerts and lack of a funding model. The false-alert problem was akin to the fable about the "boy crying wolf" too many times. Users would inadvertently push the panic button by mistake. In addition, the organization found that getting funding for apps for social good was difficult and even harder in the dramatically underfunded field of human rights. The issue of financial sustainability is ever present in tech-for-good projects across the social sector: I will definitely be discussing this challenge later!

### Successful Social Good Apps

Just because there are numerous failed social good apps doesn't mean there aren't some that have succeeded. A common characteristic of the latter apps is that they focus on essential tasks in a narrow context in which the app's particular advantages are strong enough to overcome the general disadvantages of apps. They also do three things well:

1. Give potential users a compelling reason to download the app.
2. Do something useful that users want or need significantly better than existing alternatives.
3. Motivate users to use the app regularly so its envisioned benefits are realized.

Types of apps that meet these criteria include:

**Apps where security or privacy is paramount to the user.** For example, an investigative journalist might use Signal, an especially secure communications app, to talk to whistleblowers and other confidential sources. Rather than talking to large numbers of people, where the mass communications apps such as WhatsApp shine, the app's users can talk securely with just a few specific people who share the motivation for security or privacy.

**Apps mandated or required by an employer.** If someone needs to use an app mandated by their employer to get their paycheck, that's a strong motivation to have that app. Or an app designed for community health workers to document their fieldwork will get used. Apps for this specific need are so popular that the field banded together, led by the nonprofit tech organization Medic, to create the Community Health Toolkit to make it easier to build high-quality apps.

**Apps with compelling benefits.** A student who is visually impaired is more likely to use a talking educational app for reading, such as Voice Dream's Reader or Benetech's Bookshare Reader several times a week for school. Similarly, an adult worker aiming to increase their skills to get a better job is also likely to use an app made for that purpose.

**Apps designed to work in offline environments.** Some potential users of apps lack continuous access to the internet either because of a gap in their network, a poor connection, or expense. They need apps that can work offline when internet access is not available. Kobo Toolbox, the leading app used to collect data in the field by development nonprofits and groups such as the United Nations Development Programme (the largest aid organization in the world), is one good example. Another is CoMapeo from Awana Digital (formerly Digital Democracy), which was built for Indigenous communities to collect data about their territories and enables them to keep that data to themselves.[5]

One successful app was Verto (the word *voter* rearranged), a voter advice app that founder Mike Sani modeled after dating apps, where the user would swipe right or left on policies they liked or didn't like. After they expressed their preferences, the app would share how these policies were decided by elected representatives, explain which UK parties best aligned with the user's positions, and encourage young people to register to vote. Within three years, close to two million young British voters registered because of Verto.[6]

Despite these successes, I strongly encourage any nonprofit thinking of building an app to reconsider. An outstanding guide for practitioners in civic tech on the topic of apps for social good sums up the advice by the South African activist leader Luke Jordan in its title: *Don't Build It!*[7] Jordan's point is that unless there is a compelling reason for the target users to use an app regularly, money spent on creating an app will be wasted.

Speaking of money and apps, there is usually a cheaper way to provide functionality to users than by building a mobile app. One example is a progressive web app, which is just a slightly adjusted version of a web page that can work on different kinds of smartphones. These apps are less expensive to build and maintain. They are also good for testing functionality with users. If a progressive web app catches fire with a user community, it might be a sign that users might download and use a more advanced mobile app. However, there may just be a better way to address the social need than by building any type of app.

**Nobody Will Download That App! How Bad Advice from a Management Consulting Firm Led to Figuring Out the Best Way to Modernize the Child Helpline Movement**

Apps are a very easy strategy to suggest—even if it isn't clear why or whether they make sense! In 2018, I was approached by Jeroo Billimoria, the founder of Childline India and Child Helpline International. She was trying to help the entire child helpline movement, which takes tens of millions of calls about and from children in crisis each year. A leading management consulting company had strongly recommended the building of an app for child helplines. Jeroo asked me for my opinion, and I didn't mince words: "Nobody will download that app!" I told her, "There's a reason why people memorize emergency numbers, such as 911 in the United States and 112 in much of Europe, rather than use an app to call for help."

Jeroo convinced me to look at how better tech could assist child helplines. Joan Mellea, the cofounder of Tech Matters, and I interviewed the heads of twenty-five national child helplines to figure out what they really needed. Five of these national helplines had built their own app, but requests for help made through the app represented less than 5 percent of each nation's helpline call volume. For these country's communities, the app simply wasn't the best tool for the job of assisting kids in crisis.

Of course, when I explain why a given tech idea is a bad idea, I often get asked what would be better. It is hard to figure that out without talking to the people who would use such tech. After talking to twenty-five helplines, I thought that their needs would be better addressed by building a more powerful contact center platform for helpline counselors that would add texting and social media capabilities to the emergency telephone number. Each one of the twenty-five helplines had created its own contact center solution, typically by paying consultants to heavily modify a commercial call center platform that was designed for sales teams. Wouldn't it be better to build one platform that focused solely on the needs of all nonprofit helplines that did counseling and referrals?

As a result of Jeroo's request, I put together a team at Tech Matters to create Aselo, a cloud-based crisis response platform that we designed in partnership with ten national child helplines. By adding text channels such as webchat, WhatsApp, Instagram, and Facebook Messenger, we made it possible for the helplines to communicate with young people in the channels where they chat on a daily basis.

It's hard not to notice that teenagers do much more texting than phone calling today. Beyond being a better fit for young people, text-based communications offer the opportunity to greatly improve the efficiency of counselors. It's easy to ask what language the texter speaks and route them automatically to a counselor who speaks their language. It's easy to set up a chatbot that answers important questions, just as Praekelt.org did for information about COVID. The pandemic was one of the top reasons young people reached out to the child helplines in 2020.

As of 2025, Aselo is now deployed nationally in fifteen countries, including Zambia, Zimbabwe, Thailand, New Zealand, Hungary, India, Canada, Chile, and Colombia, and is expanding to more countries. After three years of Aselo being in operation, we still haven't been asked to build an app.

## BAD IDEA 2: THE ONE TRUE LIST

Imagine a meeting taking place in a nonprofit's conference room or on Zoom. The purpose is to brainstorm ideas for applying tech to solve a social problem. Someone says, "I know! Let's make a list! Not just any list, but a definitive, comprehensive, not-available-anywhere-else list that we can put online, and people can search. The One True List!"

The excitement in the room is palpable. Everyone loves lists! The list could be an online directory of everyone working in a particular field, all the tools created for a specific purpose, all the social services organizations in a city or region, or all the grants made by foundations or government agencies for specific types of projects in country X. The One True List might elsewhere be called a "field-wide database" or a "comprehensive assessment" or a "clearinghouse": it goes by many labels. There's only one problem, though: the One True List is a generally bad idea, for two main reasons.

First, the era of the One True List is over. In pre-internet days, there were many different variations on the One True List. The idea was that a small set of printed books available in local libraries might well hold all the information anyone would need in their work and daily lives. A telephone directory was the One True List of everybody's phone number in a city or area. The advertising section of the directory, known as the Yellow Pages in the United States, was the One True List of businesses in a town. Dictionaries had all the best words. *Who's Who* books specialized in lists of important people. Trade associations published directories of members. Private publishers created directories of colleges and universities, employers in an area, and magazines and periodicals published in various fields.

The internet changed all that. Early websites focused on making lists as well. Yahoo started that way, but we don't use Yahoo anymore because Google's search engine killed Yahoo and most One True Lists. Now, people who want to know something just type keywords in a search bar or ask Alexa, Siri, or Google. The answer is served up in a fraction of the time it took to find and look through a paper list. Not only do people no longer need to know where lists are, but they also don't even need lists anymore.

The second reason the One True List is a bad idea is that nonprofits begin list-making projects based on untested assumptions that are likely wrong. Much time and money are often wasted before these assumptions are shown to be incorrect. Here are the most common false assumptions.

**False assumption 1.** Building the list will be easy because of crowdsourcing—the practice of obtaining information from large numbers of people online. Whether accomplished through direct contact or crowdsourcing, this approach depends on people's willingness to voluntarily provide information and data. The alternative is to hire an army of interns to conduct surveys or do research, which often starts by mining past list-making projects!

Crowdsourcing to create a list might sound like a great idea, but in reality few people collaborate with list-making efforts for the following reasons:

- It's been done before; it didn't work.
- It's been done before, and it does work (less common, but it can still be true!).

- It takes time away from their work, and they can't be bothered to respond with the needed information.
- It runs counter to their interests to provide data for free to a potential competitor.
- It runs counter to their mission obligations to ensure privacy and security.

**False assumption 2**: If we put the list on our website, many, many people will come and use it. This is the classic "if we build it, they will come" mentality, which has been repeatedly proven not to work by both for-profit and nonprofit organizations. The reasons are many:

- Information on a website is useless if people don't know it's there. Unless the website already has a steady stream of visitors, the nonprofit must publicize the presence of a list and entice people to come—a huge and expensive effort.
- The information in the list is perceived as much more easily accessible through a search engine.
- The information isn't comprehensive, useful, or interesting enough to engage people.
- The information is of mixed quality. The first time someone uses the list and doesn't find the answer they are looking for or, worse, encounters incorrect data, they won't come back.
- The website's tools for searching and accessing information aren't user-friendly.

**False assumption 3**: Once we create the list and put it on our website, we can relax. However, even if a comprehensive list is created and people regularly come to a website and use it, that's just the beginning. If a One True List is not well maintained, it rapidly decays and becomes useless. Keeping the list accurate and current is a huge and costly undertaking.

But here's the most important point. Even if a tech nonprofit overcomes all the development, marketing, and maintenance hurdles, the One True List is still likely to fail at its objective of achieving social good. I speak from personal experience.

### Failed Attempts to Build the One True List

I started my tech-for-good career in the assistive technology field. Assistive technology is used by people with disabilities to perform functions they might not otherwise be able to perform. Examples include text-to-speech software for people who are visually impaired, listening systems for people who are deaf or hard of hearing, and many, many more.

Assistive technology is of great interest to government agencies, including the US Department of Education. The department gave a good-sized grant to a nonprofit to build and maintain the AbleData resource database to list all available assistive technology solutions. At first, many people used the list, especially once it was put on the web. But over time they stopped using it because the information was not current or a link to a specific provider of an assistive technology no longer worked because the company had gone out of business. The list was reasonably accurate when it was launched, but the task of maintaining it was monumental. Finally, AbleData shut down in 2020 when the government stopped funding it.[8]

Another example involves apps made for use in the international human rights field. One of the biggest areas of need in human rights is being able to track numbers with respect to who did what to whom. For example, common questions are "How many people were (or have been) killed in a civil war?" and "Who killed them?" The magnitude of the conflict and who is responsible for what acts are crucial because governments take actions such as imposing no-fly zones, creating protected enclaves, and enacting arms embargos based on these numbers. The questions are equally relevant after a war has ended as societies ideally go through justice and reconciliation processes.

It is not hard to conclude that these types of lists are very difficult to create and that their accuracy is hard to verify. In fact, it's common for multiple, competing lists to exist, each one declaring that it's the One True List and that others should be ignored. For example, the Iraq Body Count, which maintained the world's largest public database of violent civilian deaths since the invasion of Iraq by the United States and coalition forces in 2003, has been criticized both for overcounting and undercounting the victims of the Iraq War.

Dr. Patrick Ball, the founder of the Human Rights Data Analysis Group, contends that these types of lists almost always undercount. To overcome these shortcomings, Ball and his team have developed techniques for estimating the total numbers of victims, including those not on any list, by statistically examining the overlaps among multiple smaller lists or surveys rather than trying to create the One True List to rule them all.

In most large-scale civil conflicts, these estimates are often the best option we have to understand the scale of mass killings. A notable example confirmed the value of this approach. Ball and coauthors used this approach to estimate that ten thousand Kosovar Albanian civilians were killed in 1999 at the height of the Kosovo conflict in the Balkans, an estimate submitted to the International Criminal Tribunal for the Former Yugoslavia. At the time, less than half of this number of victims had been identified. A decade later, a large effort was mounted to document all who were killed. The actual number came in quite close to the estimate used in the war crimes prosecutions.[9]

### One True Lists That Work

As with other categories of bad ideas, a few One True Lists have succeeded in creating significant value and social impact without overburdening the tech nonprofits that created them. Here they are and why they work.

**Lists of mandated information.** The GuideStar database contains the public financial records of all US nonprofits, which by law need to be both filed with the Internal Revenue Service and made publicly available. These records are updated annually because these returns must be filed each year. As a result, GuideStar gets used frequently when people are looking for details about a specific nonprofit or funder. GuideStar also has come up with information products to sell, which creates revenue for its nonprofit activities. It uses a business model known as "freemium." For example, rather than painstakingly looking through dozens or hundreds of returns to find the salary information for a specific position, it's possible to buy a salary report that makes it easy to find the salary ranges for, say, a chief financial officer working for a nonprofit with an annual budget between $10 million and $25 million. Many nonprofits willingly invest the time it takes to improve their GuideStar listing because they know it will be looked at by donors considering a grant to them. GuideStar also encourages nonprofits to add more data than is contained in the standard government returns by awarding them with Silver, Gold, and Platinum badges. So not only is most of the information on GuideStar mandated, but the organization has also created a powerful incentive for the crowdsourcing of additional information from nonprofits (beyond what is in the annual required form) hoping to enhance their chances of getting funded.

**Network-powered lists.** Lists made by a commercial company with the power of a huge network behind it, such as Facebook, LinkedIn, or Google Maps, will almost always clobber a list created by a small organization. Three rare exceptions are Wikipedia, the Wayback Machine, and Open Supply Hub. Wikipedia, the massive online encyclopedia, managed to get the power of a volunteer network to crowdsource the creation of one of the most used websites on the internet. That amount of usage led to funding appeals for donations that generate more than $100 million annually to keep Wikipedia maintained and improving.[10]

The Wayback Machine, created by the Internet Archive, uses advanced storage technology created internally to capture and archive copies of webpages and makes them available to people for free. If you want to see a given website as it appeared one, five, or more years ago, the Wayback Machine is likely to have a snapshot saved. It's an essential resource for me, especially when doing research and writing this book!

Open Supply Hub has built a list of the facilities used in the global supply chain. Its founder, Natalie Grillon, was inspired to tackle this problem by the Rana Plaza disaster, in which more than one thousand Bangladeshi textile workers making products for major western brands died when their workplace collapsed.[11] Open Supply Hub creates a common standard for keeping track of facilities' workplace safety, working conditions, and environmental footprint. Because the reputation of major companies and brands depends on the data and standard identifiers on Open Supply Hub, that creates an incentive for many stakeholders to ensure that the database is kept up-to-date.

**Lists of federated data.** These lists expose data in a known format, making it easy for many different players to provide access to comprehensive information. For example, there is no one global list of restaurants, but there is a web standard for restaurant information that is generally followed by the restaurants themselves (on their web pages). This makes the information easy to federate if you're Google or Bing or TripAdvisor or Yelp.

Building big lists requires a big investment of time, energy, and money and even more effort to maintain. It is thus a big risk for most tech nonprofits to take on the chance that people might use it. For almost any social problem that urgently needs to be solved, there is likely a more effective way to address it than building the One True List.

## BAD IDEA 3: "THE NEXT BIG THING" IN TECH

New technology arrives on the scene all the time, promising to revolutionize how creators build technology products and users use them. Some are the real deal and break through into mainstream applications and widespread adoption (think cloud computing). Others hover on the edges before arriving in successive waves of innovation (think AI and machine learning). Still others settle into a niche, with utility in a narrow set of applications (think virtual reality, despite Facebook renaming itself Meta, short for the virtual reality metaverse). Finally, the jury is out about the utility of some heavily hyped tech (think blockchain).

However new technology in all these categories ends up, it is often initially hyped as "the next big thing in tech" and massively oversold. This pattern is well documented and tracked in a model called the Gartner Hype Cycle.[12] Developed by a leading tech research firm, the model helps people take a more measured approach to new and exciting technologies. Jake Porway, the founder of DataKind and a leader in applying data to social good, pointed out in a Good Tech Fest speech entitled "Breaking the Cycle" in 2024 that tech hype cycles are happening more and more frequently and that social sector leaders seem to fall for them each time.[13]

Which leads to the next bad idea in tech for good: choosing a technology simply because it's "hot" rather than because it's the best technology for the job. Over the past decade, blockchain has been a near-perfect example of an overhyped technology. More recently, generative AI platforms such as ChatGPT are the latest example of this pattern. However, AI is one of those technologies that has been through multiple fad cycles and comes out stronger (and more useful) in the end. AI is such an important topic in tech that I've devoted an entire chapter to the subject (see chapter 5, "Building Successful AI Solutions").

Blockchain is also an ideal example of how an exciting technology frequently isn't ready for nonprofit use. Blockchain emerged in the early 1990s as a highly secure ledger (database) that made time-stamped records difficult to tamper with. The popular vision of blockchain as a distributed, shared ledger across a network of participants where everyone has access to information but no one has control was described in 2008. The magic of a blockchain database is that a group of transactions is bundled into a block and then that block is locked to a chain of blocks stretching back in time. The math behind blockchain makes it impossible to change any of the past blocks (including the transactions in them) without the tampering being obvious to everybody.

These attributes make it ideal for recording transactions and tracking assets in low-trust environments. A blockchain ledger can be shared across public or private networks. Public blockchain networks are used to support digital currencies such as Bitcoin, launched in 2009: anyone can have a Bitcoin wallet, but there is no need to know who they are. Private blockchain networks are more suited to banking and financial services, where it's essential to know exactly who is on the network. Blockchain is a very specialized kind of ledger. However, as blockchain gained notoriety in the 2010s, it was touted as breakthrough technology that could be of benefit in any type of transaction.

In the nonprofit world, major donors started putting out requests for proposals for projects using blockchain. This practice continues even now, when—surprise!—it has become obvious that blockchain is almost always a terrible idea for social good applications. The typical exaggerated claim in the social good field claims in effect that blockchain can solve intractable human problems with *math*. Math is great. I love math! However, tough social problems are linked with pervasive patterns of human behavior. Math alone doesn't have the power to change people's behavior. These miracle-cure claims are unfortunately implausible in the real world. In addition, blockchain is difficult to set up, expensive to use (often requiring enormous amounts of energy to run), and slower than most databases. Like most of the "Next Big Things" in tech, it might end up being useful for some narrow niche applications, but it's not a panacea.

The excitement around a fad generates bad ideas that can take a long time to fade away. Areas in which social good tech applications using blockchain have been tried but failed to meet expectations include:

**Land rights.** One of the most common use cases cited for the wonders of blockchain is in land registries. The assumption is that land title records in a blockchain database would prevent corruption. Unfortunately, issues of land rights are tremendously complex. By itself, a blockchain database does very little to address the lack of respect for the land rights of disadvantaged people. A corrupt elite that controls the land titling system and regularly appropriates lands from Indigenous and impoverished communities will simply ignore the wonders of math in the form of an alternative blockchain-driven land record system. Or it might simply cement historically unjust land ownership patterns and confirm past misappropriations.

**Payments.** Blockchain was supposed to revolutionize payments in the social impact sector. More than five years ago, the World Food Program claimed that using the Ethereum blockchain in certain refugee camps in the Middle East reduced its bank transaction fees by 98 percent.[14] With an impact like that, why hasn't *all* provision of cash benefits in global development moved into blockchain? Probably because the other costs involved with using Ethereum as a payment mechanism in practice exceeded whatever savings there might have been in reducing bank fees.

**Decentralized autonomous organizations and smart contracts.** There was early fervor in the blockchain world for decentralized autonomous organizations (DAOs) supported by smart contracts. A DAO is a collectively owned, blockchain-governed organization without leaders. Smart contracts define the rules of the organization and hold its resources in the form of a cryptocurrency. There is no centralized authority: decisions are typically made by a vote of the members and the software decides when a vote has been approved or if a certain action is permitted under the rules. The benefit of a DAO in a low-trust environment is that its members can count on software instead of a human to do the right thing and make payments, or change policies that affect financial outcomes, appropriately.

Imagine big piles of money online in a blockchain run by software that hands out money when certain conditions are met, all without human oversight. Control of money, which typically resides in a bank or with an escrow agent, shifts to a piece of software. What could possibly go wrong? As it turns out, plenty. An early DAO launched on the Ethereum blockchain in 2016 was designed to act as an investor-directed venture capital firm. The DAO raised $150 million in ether, a digital currency. Three months later the DAO was hacked, and $60 million of ether was stolen.[15] More crypto hacks followed against similar projects; more money was wasted.[16] Betting millions that software will have no bugs turns out to be a bad bet.

### Where Blockchain Might Work in Social Good Applications

Unlike for the other bad ideas in this chapter, I was unable to find the 5 percent positive exceptions in the blockchain area. As of 2025, I couldn't find any social good application of blockchain technology operating successfully at scale. The technology is simply at too early a stage. I do think that blockchain has some intriguing attributes and might have success in the future. However, the typical nonprofit organization should stay well away from it for now!

The social good blockchain opportunities that might be successful are likely to be in high-value, low-transaction-volume applications where the technical advantages of blockchain shine. One application that has promise is in the supply chain field, specifically in ensuring an ethical and sustainable supply chain.

Now, you might wonder why nonprofits care about the supply chain, given that it is a business construct. It turns out that many social problems come about as a direct result of bad behavior by businesses in the supply chain, such as:

- Child and slave labor
- Unsafe working conditions
- Deforestation
- Poaching
- Illegal fishing and overfishing
- Air, water, and land pollution
- Excessive use of plastics and overpackaging
- Excessive use of antibiotics and inhumane conditions in the dairy, poultry, and meat industries
- Unsustainable use of water resources in the face of the climate crisis

That is just a sampling of why the nonprofit sector sees changing the behavior of supply chain actors as one of the top ways of accomplishing their social justice or environmental mission goals. So how can nonprofits create incentives and disincentives that might shift the supply chain away from bad practices to good ones?

It turns out that a growing number of consumers today care about the products they buy and would like to know if the social claims that companies make are true. If a product is presented as being from a woman-owned business, created without animal testing, or offsetting its greenhouse gases, can consumers believe that claim? If they know with greater confidence that the claim is true, it's reasonable to assume that the product will have greater sales and might even be able to justify a premium price.

Two for-profit blockchain tech enterprises, Provenance and OpenSC, are working on this challenge. The Working Capital Fund, an impact investor dedicated to social impact while making money, has invested in both of these companies: its idea as a social investor is that it can go to a for-profit company to raise venture capital from regular investors and convince the company to add social impact elements into their commercial product. Both are examples of companies focusing on building real products for real consumers and brands. (Full disclosure: I advised the Working Capital Fund.)

I met the founder of Provenance, Jessi Baker, during the company's early years. Baker paid some serious technical dues in the blockchain technology world while working on a computer science PhD at University College, London, and had a vision of blockchain as revolutionizing the supply chain. Her idea was that a public blockchain could be used to track products every step of the way along the supply chain—for example, from a fishing boat in the Indian Ocean to the seafood case at the local Waitrose supermarket in London.

Ten years into her entrepreneurial journey, Baker is leading a good-sized company building products for major consumer brands. Her vision of blockchain's potential hasn't changed, but her experience in working with the reality of consumers, large companies, and the supply chain has shifted what her team is focusing on. Rather than trying to touch every step of the supply chain, which turns out to be incredibly resistant to change, Provenance is concentrating on confirming claims about ethics and sustainability. It is creating blockchain-based ways for consumers to confirm that these claims about third-party certification are true.

When a technology is still in its early days, as blockchain still is, I believe the social sector should leave the expensive experimentation to brilliant technologists and entrepreneurs such as Jessi Baker. Let the big-money people fund this phase and surface the most useful applications of the technology, which paves the way for tech nonprofits to scoop up that technology later and borrow it for cheap!

Will there ever be broad adoption of blockchain in social good applications at scale? I am certain there will be, but right now it's too early to tell where and when. In the meantime, an easy test for blockchain claims in the social good space is the substitution rule. If you replace the word *blockchain* with *database*, does the claim cease to make sense? If so, the blockchain claim probably doesn't make sense.

I have gone deep on blockchain not because it is the only overhyped technology. The word-substitution trick works well in assessing other new technology. I like "spellchecker on steroids" instead of "generative AI" for the current generation of generative text AI tools such as ChatGPT. The only thing I can be sure of is that something else will be coming up as the

Next Big Thing every few years. The main thing about the Next Big Thing is not to lose touch with the basics of applying technology successfully, such as solving a real problem for humans first rather than trying to find an application for the latest tech fad. When someone promises a miracle cure for a social problem by means of the Next Big Thing in technology, we should be more than a little skeptical.

### BAD IDEA 4: THE CULT OF THE CUSTOM

The Cult of the Custom springs from the belief, often sincerely held, that your nonprofit is unlike all other nonprofits, so it needs a customized software solution to meet its needs even though it has an annual budget of less than a million dollars. This bad idea rarely ends well.

Imagine if every restaurant thought it needed its own software! In the real world, restaurants have many options for software, each used by thousands of restaurants to help them track orders and charge a credit card. Although the software might not be perfect, restaurants realize that customers don't care about the software used by the staff. Customers care about things such as food quality and service. So restaurants pick software that works well enough for them and focus on the things that matter to the success of their business. The same is true for other businesses with their own groups of target customers—for example, dental offices and golf courses.

This doesn't mean that there aren't differences among restaurants, dental offices, or golf courses and that it wouldn't be good to have some customized features. But software vendors can't rewrite their software wholesale to work exactly how each business thinks the software should work, so they typically build enough flexibility into their products to accommodate most of the differences.

The Cult of the Custom in the nonprofit world thrives on bad advice. When struggling with technology, nontechnical social good leaders often turn to tech consultants. These consulting businesses serve as intermediaries between tech companies and their customers and provide a range of services from implementing, configuring, and customizing standard software to building software from scratch. When a nonprofit describes its challenge, too many tech consultants are quick to recommend building custom software from scratch because there is more money and profit for them in doing so. Or they might suggest customizing a platform that was designed for large organizations they know well, such as Salesforce. Maybe they don't even know that a better option exists, but they have few incentives to explore a better option.

Even if the tech consultants are volunteers, and you aren't paying them, the situation is still the same. The nonprofit world is full of volunteer-created solutions built with the best of intentions but ultimately unsustainable. Free can sometimes be very costly, especially if your team wastes hundreds or thousands of hours on a tech solution that doesn't end up working.

Nonprofits can receive equally bad advice from management consultants who, no matter how good they are at providing strategic advice, are generally terrible at making recommendations about technology.

Social good leaders, beware! Custom software is almost always the wrong choice for the small nonprofit, especially if it has no internal tech expertise. Here's why:

- Software development is expensive, and custom projects require more development compared to minor adaptations to standard platforms. It's not uncommon for a custom project to expend its full budget (or even a whopping 200 percent of the budget) without getting to a fully functioning system!
- Dependence on outside developers is risky. I have encountered numerous nonprofits where the sole developer of their custom software has abandoned them for a better-paying job, leaving the nonprofit with half-finished software. The main financial interest of tech development firms is in selling new software projects, not in maintaining past ones.
- Software needs constant updating, whether for bugs, security risks, or feature enhancements. As the only customer for custom software, the nonprofit has to pay for all of that.

Building custom software is not a bad idea when a nonprofit has strong tech expertise and a proven track record of building software. There are many terrific examples of tech nonprofits, even though the outside world might not perceive them as such. For example, Kiva is a well-known microcredit nonprofit, but if you peek under the hood at Kiva, you'll see a software company in operation. Benetech, the tech-for-good nonprofit I founded and led for thirty years, has a software development team with a budget in the millions of dollars a year. Even Callisto, a nonprofit that focuses on detecting serial sexual assault perpetrators on college campuses by empowering survivors, bases its promises to survivors on its tech solutions.

Some nonprofits have also found themselves strong technical partners to build solid custom software with them. For example, the Swedish child helpline, BRIS (Barnens Rätt i Samhället, or Children's Rights in Society), years ago created a special tool that takes all text conversations the organization has with children and scrubs them for personally identifiable information—such as proper names and place-names. This helps BRIS both in maintaining compliance with privacy laws (the General Data Protection Regulation, or GDPR, in Europe is currently the most stringent such regulation in the world) and in keeping its commitment to being an anonymous helpline.

The scrubbing of personal data also allows the helpline to keep and analyze the anonymized text conversations, which could be useful in seeing patterns in conversations over time. BRIS was not going to market this tool, in part because the tool works in Swedish only but for the most part because BRIS is not in the business of selling software. However, new privacy laws such as the GDPR made doing something like this a great idea: many organizations needed it. The market later decided it was a worthwhile product. It's now available from multiple commercial software vendors such as Amazon Web Services. However, it's not available in most of the world's languages yet.

Another example is VillageReach, the Skoll Award–winning nonprofit that focuses on bringing health innovations to the "last mile" in rural Africa. Because it needed to manage its own warehouses and medical supply chain, VillageReach created logistics management information system (LMIS) software. Other nonprofits and government health agencies needed it as well, so VillageReach decided to build the software as an open-source project. OpenLMIS is now widely used throughout the public health arena in Africa. VillageReach found it challenging to be both an innovative health program and a software enterprise, and eventually spun the OpenLMIS project to a separate organization that continues to maintain it.

BRIS and VillageReach are rare examples of the 5 percent of the time that custom software makes sense, but these nonprofits had access to tech savvy that most nonprofits don't.

### CONCLUSION

The bad ideas I have just shared are not always terrible ideas; almost all of them have been applied successfully somewhere. However, social change leaders receive persistent messages that they should be able to succeed with a particular technology, even if that tech has been overhyped and many other people have failed to apply that innovation successfully.

Tech ideas are hard to pull off, especially when you assume a particular technology is the right one. Many tech ideas that seem as if they should work don't work at all. As I share in the next two chapters, there are design approaches to solving real problems that tend to fail much less often.

By sharing stories of tech gone wrong, I hope to make it easier for social good leaders to understand their own limits and avoid making the same mistakes as those who have come before them.

The single most important thing is to ensure that people are at the center of any tech initiative: the people who are being served by it; the people who will use it; the people who are designing and building it; and the people who improve and maintain it. Looking to their needs and underlying issues with a keen awareness of the limitations of technology raises the odds of success dramatically higher compared to starting with a tech solution in the false hope that it is certain to help.

# 3 THE THREE BEST TECH-FOR-GOOD IDEAS

I hope you're not daunted by the litany of bad ideas in the preceding chapter! Failure in tech is normal. It's how we learn what works. The magic is to fail early and cheaply, thus increasing the chances of breaking the code for success before you run out of money.

Just as we now know a great deal about what doesn't work in tech, we also know there are patterns and approaches that do work. They are worth keeping in mind as you try to do amazing things for society. I'll warn you: they aren't necessarily easy, and they frequently go against ingrained habits in the social good sector. For example, being effective with tech almost always involves changing human behavior. The benefits of the new tech must be sufficiently strong to overcome ingrained habits that will need to be discarded or adjusted for the tech to be successful. Luckily, we have a great deal of data that show tech is reasonably good for driving behavior change.

My first pattern for success in tech for good is to equip the people on the front lines with better technology: this is where social change actually happens and where behavior change makes the most difference. Next, you need to kill the dinosaurs—the old approaches and organizations that used to be innovative but have outlived their usefulness. And then you must aim for the clouds when creating software that collects much more useful data and delivers better results.

The three good ideas for tech described in this chapter share a common theme: the tech is deeply embedded in the program activities of the organization. I can't promise you that using these good ideas will guarantee you success 95 percent of the time as the virtuous opposite of the 95 percent rate of bad ideas in the previous chapter. However, most successful tech-for-good organizations I have encountered have embraced at least one of these ideas and frequently all three.

## GOOD IDEA 1: EQUIP THE PEOPLE ON THE FRONT LINES

Modern for-profit organizations realize major productivity gains by giving their employees powerful tech tools. It's also common for them to provide easy-to-use tools for customers. A business saves a lot of money if a customer can order a product, pay for it, install it, and solve any problems with the product without ever needing to talk to an employee of the business. As a result, businesses today are highly focused on figuring out how their customers and their employees can get more and more done in less time and with fewer complications. And if everything can be done online without the need for a physical store or employees on the ground, even better.

As noted earlier, nonprofits are often not very good at equipping their teams with current tools. The pandemic drove home to many nonprofit leaders the impact of failing to do so: layoffs, shutdowns, and the shortchanging of the communities served. During the pandemic, there was a huge pivot to providing services online when that became the only practical option to continue the work. After the pandemic, that pivot opened the door to more and better technology in the social good sector.

Some of the most exciting tech-for-good enterprises of the next decade will focus on building great tools for frontline workers and community members. I recommend focusing on and listening to these two groups of people both because they have the greatest need and because there are many more of them than there are senior organization executives. Tech is best deployed where enough people have enough common needs to justify creating a new product or platform. Plus, having many more people using your product generates more data that can be used to improve the product and serve the users better. Helping more people and helping them better should be the guiding principles of tech for good.

If you step back and think about social good work, most of it is based on intangibles. Yes, someone who has been hit by a natural disaster needs physical items such as shelter, clean water, and food. However, so many of the activities delivered by nonprofits and agencies are fundamentally about collecting, processing, and providing information. Classic examples include training and education, assessing which programs a person qualifies for or needs, and tracking the efficacy of a supply chain. These tasks are well matched to software and data solutions, like the dozens discussed in this book. Every nonprofit that actively pushes around information is ripe for the right kind of software to improve effectiveness.

That's not to say that all programs based on humans interacting with each other should be replaced with technology. Unlike the for-profit sector, the social sector prioritizes human contact. But, given that the social sector never has enough time or money to respond to everybody in need, it is usually worth exploring which activities can benefit from more technology and more automation. If you have the same employee engaging in both high-value personal interactions and low-value activities, you should automate some of the low-value activities to free up the most valuable asset in any organization: your people.

Even though the major impact of tech innovation should be on the front lines of the work, the needs of organizational leaders also have to be addressed. As a secondary priority, data from program delivery should be molded into useful dashboards, reports, and visualizations for managers and leaders. This is important because leaders make the decisions about investing in tech tools. The fact that the people benefiting the most from social good work, including tech innovations, are not the people making the buying decisions is part of the structure of the nonprofit sector. The best solutions shift more power to the front lines because that's where the impact lies.

Beyond providing the staff with efficiency-enhancing tech, we have an even bigger opportunity in bringing tools directly to the people in the communities we aim to serve. Many social problems can be solved by individuals with the right supports. There is tremendous leverage if you can replace an existing service that requires a staff person with a digital solution that brings these benefits directly to an individual. It is hard to understate the effectiveness of approaches that, instead of intervening on behalf of the communities we serve, shift power to them to solve their own problems.

People in the community you are attempting to serve often understand the problems they face and the context of those problems better than external actors. The entire social sector is looking at shifting away from doing social good interventions for people to instead partnering with individuals and communities to provide assets and tools for them to pursue their own goals.

One appealing example is the work done by Fundación Capital, which relied on frontline women to provide access to a financial education app by entrusting women in poor communities with a tablet with the organization's app preloaded. Fundación Capital's trust was well placed: it didn't experience equipment losses, and the application was widely used by illiterate and semiliterate women in Latin America to learn more about money, ATMs, and other topics important to their economic lives.[1]

This book is chock-full of stories about nonprofits that built tech tools to support the community members themselves as well as the staff and volunteers helping the community directly.

**Callisto: Empowering Survivors of Sexual Violence**

Survivors of sexual assault are a community nobody wants to join, but many are forced to do so. Sexual violence is a widespread problem in human society. College campuses are no exception. Twenty percent of women students report being sexually assaulted during their time in higher education.[2] Survivors have a widespread (and generally well-founded) perception that law enforcement and university administrators do not take them seriously. As a result of this broken system, roughly 90 percent of sexual assaults on campuses are committed by serial perpetrators who experience few consequences from their violent behavior.[3]

Jess Ladd is a survivor of an assault that occurred while at college and the founder of a nonprofit working to advance sexual health. She had an idea with the potential to shake up the status quo. What if a system could securely match survivors who name the same perpetrator? Jess's vision led her to create Callisto, a Skoll Award–winning tech nonprofit. The Callisto system creates the opportunity for survivors to securely submit a time-stamped account of their assault. It also allows them to enter into the Callisto matching system the name, phone number, email address, and/or social media handles of the perpetrator, information available to them because more than 85 percent of sexual assaults are committed by someone known to the survivor.[4] This information is also securely encrypted, so the staff at Callisto can't read the accounts or know identities of either survivors or perpetrators.

The exciting twist in the Callisto technology happens when different survivors enter in the same information about the identity of a perpetrator. The system spots that the scrambled versions of this information are the same. For example, if the phone number matches, the Callisto team doesn't know the phone number, just that the scrambled versions of the phone number are the same. Then the two survivors are offered the chance to speak with a legal options counselor. Callisto is focused on survivor empowerment, so it simply offers this option. The survivor is free to accept or ignore the suggestion.

Survivors often choose not to report sexual assaults because of the common knowledge that this path is painful and typically fruitless. However, Callisto has reason to believe that when two or more unrelated individuals name the same perpetrator, the power dynamics change. First, multiple survivors accusing the same perpetrator are much more compelling in today's system than a one-on-one dispute of the "he said/she said" or "I said/you said" variety (it is not always a case of a male assaulting a female). Second, survivors who might not go public on their own account may instead choose to risk the consequences of doing so out of solidarity with future possible victims of a perpetrator who seems to enjoy impunity for their actions.

Human rights innovations generally face sustainability challenges, and Callisto is no exception. Survivors are not in the market for these kinds of services, and charging for the service would undercut the fundamental social justice model. Callisto initially partnered with the Title IX offices on US campuses, which are named for the federal educational rights law that forbids discrimination on the basis of sex. Although some Title IX offices were terrific partners (and willing to pay to bring Callisto to their campuses), it became clear that others were not excited about improving the reportage of sexual assault on their campuses. Callisto ended up abandoning this approach (and foregoing the revenue opportunity), deciding instead to go directly to survivors and have legal options counselors work with each survivor. It needs philanthropy to underwrite its work because its users are not in a position to pay for its survivor-centric supports.

This dependence on philanthropy, common to almost all civil and human rights organizations, is risky. In late 2024, Callisto announced it would shut down because of insufficient funding. Luckily, a number of donors stepped up with longer-term funding and Callisto was able to resume operations within two months.

By equipping survivors with the tools to securely capture information about their assault and their perpetrator, Callisto gives survivors more information and more control over what happens to them. Callisto has had numerous matches on the first forty college campuses where its latest tool was deployed, and in 2023 it expanded to all college campuses in the United States. Full disclosure: I was impressed enough with Callisto that I served on its board of directors for six years.

### Touching Hundreds of Millions

Equipping the front lines is a recipe for social good at a massive scale if the problem being solved is common to hundreds of thousands or millions of people. Two organizations exemplify this kind of scale: Kobo and Medic. Kobo Toolbox set out to equip frontline survey takers working on disaster response and other humanitarian crises with a free tool that made it easy to collect data about needs in the field. It's designed to work in many languages and collect many types of data, even if the survey respondent is unable to read or write or has no internet connectivity. In 2022, more than two hundred million survey responses were received by Kobo's servers.

Medic focuses on meeting the needs of community health workers. The community health worker model was pioneered by the late Paul Farmer, founder of Partners in Health. As a physician, Farmer knew that most humans' medical needs would not be addressed by a doctor anytime soon. By using trained community health workers, however, it would be possible to extend more and more health care cost-effectively to poor communities around the globe. These health workers needed much better technology to do their work, so Medic's goals were both to replace paper forms with digital ones as well as to connect the community health workers to the larger health-care system in each district where it is used. Medic's main product is an app for these health workers to plan their visits to families and to capture critical health information. It is a great example of the 5 percent of apps for good that make sense! But the focus is not on the hospitals or the doctors: it is on the community health workers who work with local community members to help mothers give birth safely, to track and treat diseases, and to ensure children are vaccinated.

As I shared in chapter 1, Benetech's first product was the AI-equipped Arkenstone reading systems for people who were blind or visually impaired. At the time we founded the organization, the status quo was that blind people were read to. Sighted readers often had opinions about what these people should and should not be reading and made editorial decisions about what parts of a book or article to omit. They essentially acted as de facto gatekeepers over the books blind people were permitted to read, which was quite problematic. Sighted readers were also rarely available when it was convenient for the person with the disability: perhaps reading only on Saturday afternoons! Libraries for the blind also decided what books were worthy of recording or converting to braille. I remember one of our blind users talking about growing up hungry for any kind of braille book because he had access to so few. As a kid, he ended up reading a book on adult incontinence that was mailed automatically to him by the library for the blind because he had nothing else to read in braille.

This power dynamic was upended when this community gained access to a personal reading machine capable of recognizing the print in a book by scanning its pages. This meant that blind people, just like sighted people, could choose their own reading materials. They could choose to read a trashy novel, a political tome, or a religious work that might not be acceptable to a volunteer sighted reader. Or they could read a boring nonfiction book they needed for school or work, which sighted readers didn't enjoy reading aloud. Putting the tool of accessible reading into the hands of individual people with disabilities that impaired reading gave them agency over reading that they previously lacked. It was the beginning of a reading revolution for people with disabilities that interfered with reading, such as blindness. Suddenly, millions of books came within the reach of an individual who wanted to read them badly enough to spend a few hours scanning a book page by page.

#### Napster Meets Amazon Meets Talking Books for the Blind, but Legal: How Bookster Became Bookshare

Benetech's next social enterprise built on this power shift. Interestingly enough, the idea came from my then fourteen-year-old son Jimmy, who loves music.

It was the peak of the dot-com boom in 2000, and I came home one night and found a new program had been installed on the family PC. I yelled to Jimmy: "Haven't I told you not to install software from the internet?!" Jimmy explained that it wasn't just any software from the internet. He had gotten the software from his friend Chris, who lived two doors away. Chris's mom, Eileen Richardson, was the acting CEO of a new tech company called Napster. I had never heard of it.

Jimmy proceeded to sit down and show me the original Napster, the first peer-to-peer music-sharing system. He liked punk rock (he was in a band with Chris) and shared some punk tracks with me. I was able to find some Pat Benatar and claimed that I had some cool music back when I was young. We had a fabulous time. After an hour of time with my teenage son, which was a miracle as far as I was concerned, I was ready to pay anything for this product. However, Jimmy said that it was free and was always going to be free. I thought, "This is so illegal. But it's so cool!" And so an idea was born.

I went to Gerry Davis, the same attorney who had helped start Benetech as a non-profit and continued to serve on our board, with my new idea: Bookster! We would create a new digital library based on crowdsourcing the scanning of books using the Arkenstone (and other) reading systems. Instead of hundreds of families buying the latest Harry Potter book (at the time) and scanning it for their child who was visually impaired, one person could scan it, one or two proofread the scan, and then the library could share it with not only the hundreds of families who had been willing to spend several hours scanning a giant book on a home scanner, but also with thousands of kids whose families weren't able to afford the book, the time, or the scanner.

Gerry came back quickly with surprising news: Bookster was legal under US copyright law! It turned out that a law had been passed less than five years earlier, making it legal for nonprofits helping people with disabilities that interfered with reading print to distribute accessible versions of books without needing to get permission or paying a royalty to the copyright owner. The law covered not only formats such as audiocassette tapes and braille books but also, amazingly enough, digital text. Bookster was a rare case of coming up with a great idea that probably should have been illegal but wasn't!

Gerry did have two recommendations, though. First, he strongly suggested that we not call the new nonprofit project "Bookster" because that would freak out book publishers. Second, he suggested that we sit down with publishers well in advance of launching the library and allow them the time to work with us so that we could get their support.

We followed Gerry's suggestions. We renamed the library "Bookshare" instead, emphasizing our goal of building a library based on sharing among the blind community. Talking to the publishers early did a tremendous amount to gain their trust: they had specific concerns that we were able to address as we built the Bookshare platform. By the time we launched Bookshare, the head of the Association of American Publishers, Pat Schroeder, sent out an email to all of its members telling them that we were not an e-book piracy ring and that we were trying to help blind people, something the publishing industry had a long tradition of supporting.

The fact that more than forty thousand people with disabilities had Arkenstone reading machines created the conditions that made Bookshare possible. Bookshare became the first library to be built by people in that same community. Instead of sighted people volunteering to help blind people, a noble instinct but one fraught with uncomfortable power dynamics, blind people were volunteering to help each other. With a digital platform that enabled a powerful network effect, the sum of that community's efforts were far beyond any one person.

The idea worked! Soon, Bookshare was the largest library for the blind in the world, and initially just about every book in the library was there because a reader in that community had wanted it. They could download and be reading an accessible book one minute after they decided they wanted it. No need to get someone to help them find the book or to read it to them, and no need to spend hours scanning it.

We also saw that both the Arkenstone Reader and Bookshare were adopted by people who weren't blind. Parents and teachers of children with disabilities had often been recording books or creating braille books for those children. Having access first to a scanning system with optical character recognition and then to a giant accessible library of books saved them a great deal of time. Also, as I shared earlier, we stumbled across the fact that these products were also very helpful to people with learning disabilities such as dyslexia.

Together, these two successive innovations, the Arkenstone Reader and Bookshare, helped change the reading experience of people who have disabilities that interfere with reading printed books. By equipping those on the front lines of book accessibility, we changed the entire system for the better.

Hundreds of organizations have built tools to equip the front lines of people affected by the very diverse social issues the organizations are addressing. By asking the people at that crucial interface between community and social good organizations what their biggest problems and needs are, you might find the next big opportunity to use tech to make a difference for them.

## GOOD IDEA 2: KILL THE DINOSAURS

Large parts of the social sector are dominated by dinosaurs: nonprofits or government agencies doing their work the same way it's been done for decades. A social innovation that was amazing in the 1940s, 1960s, or 1980s has very likely run its course by the 2020s. If incumbents are resistant to change, a great approach for a smart social good leader is to use tech to scale a new, far less expensive, and far better way to accomplish the social objective of delivering social impact. The dinosaurs that don't update their approaches should eventually die of irrelevance.

In his groundbreaking book *The Innovator's Dilemma* (1997), Clayton Christensen documents how for-profit companies that have innovated in the past and achieved market dominance are often reluctant to overturn the steady stream of profits coming their way.[5] Innovation stops because the companies don't want to upset a profitable status quo. Customers still like the products very much and want to keep buying them. Highly successful companies with a flagship product have a particularly hard time innovating a highly profitable but aging product. They are susceptible to new innovators who show up with a disruptive new product that ends up displacing the incumbent product. As a result, large companies in the tech industry have learned their lesson and frequently use their financial resources to buy out new innovative companies. Sometimes they shut them down (shortly after buying them!) to extend the life of their cash-cow product (dinosaurs as cows?). Alternatively, these acquired companies sometimes breathe new life into the larger company that is in danger of losing its edge.

As noted earlier, the nonprofit sector moves even more slowly. Powerful forces maintain large incumbents in place. Nonprofits depend on donors rather than customers, which is another example of the social sector's structural problem, where the people paying are not the people benefiting from the work (whether it is technology or a program). A big part of the challenge for tech-for-good innovators is to buck this imbalance and focus on the needs of people on the front lines of social change, both the people delivering the services and those benefiting from them.

It's hard to make a major shift in how you operate your nonprofit when your donors have come to expect a very specific program from you and your brand. Nonprofits that depend on government funding also struggle to change because government contracts are usually highly specific on exactly how the nonprofit is to operate its funded programs. It's hard to innovate when the "how" you deliver your program is dictated to you rather than just the "what" of the needed social impact.

As time goes on, an organization can do everything "right" and still lose its position and fade from relevance. The dinosaurs in a given field often think they are doing the best job possible. They can confuse their program activities with their mission, which may lead the organization to thinking that its mission is to continue delivering its existing programs rather than reinventing them to better meet the needs of the people they serve. These legacy programs treat specific symptoms of a social problem in a traditional way rather than addressing its root causes.

Even after dinosaurs in the social sector become less relevant, they can go on living for a very long time thanks to donor inertia! Once certain practices are established and the donor funding keeps flowing, such organizations are hard to dislodge even long after they outlive their usefulness or their cost-effectiveness.

Kevin Starr of the Mulago Foundation, a leading funder of social entrepreneurs working on global poverty, has a different descriptor for the organizations I label "dinosaurs." In his article "Don't Feed the Zombies" (2023), he blames the funder community for supporting ineffective nonprofits.[6] He cites research from Max Roser of Our World in Data that shows that nonprofits can vary by more than one hundred to one in cost-effectiveness in solving a particular social problem. Worse, some of the evaluated nonprofits seem to be actively harming the people they are supposed to serve!

Dinosaurs are frequently filled with good people, and I am not advocating that those people suffer through job loss that may come with innovation. However, if there is a far better way to address the needs of a disadvantaged community, we owe it to the community to stop funding programs that are relatively ineffective. The people working for dinosaurs could be better deployed doing something much better. Many of them know this but don't know what to do instead.

You might work in an organization with a legacy program that bears a startling resemblance to a dinosaur. Perhaps you have begun thinking about how to innovate with a more modern and better approach to the same social problem.

It is easy to think of a large bureaucratic organization as a dinosaur that needs to be replaced. Many social entrepreneurs started their careers to overcome failings they saw in government agencies or larger nonprofits. However, over the long arc of a successful social enterprise, the dinosaur we must kill might be our own! Rather than waiting for someone else to kill the dinosaur and replace you, better to pivot to a better way of delivering on your mission. Pivoting is also less violent in that it refashions your aging dinosaur of a program into something new.

Yesterday's insurgent can easily become today's dinosaur. Sometimes you need to keep the program operating in a similar way but reinvent the technology. Sometimes you need to move higher up in the social good food chain by shifting from direct services to influencing others in the field to update how they operate.

Quite a number of successful social entrepreneurs founded pioneering organizations that provide direct services and then moved into field-building roles where they focused on helping other organizations, including those they might have seen as competitors in their earlier stages. These social entrepreneurs have transcended the day-to-day competition among nonprofits for funding and realize the entire field would be better off by adopting the innovation they pioneered.

Mothers2mothers is a nongovernmental organization, or NGO (the international term for nonprofit), founded in South Africa. I very much admire the pivot it made. Its initial innovation was to connect pregnant mothers who have HIV with mothers who are HIV positive but successfully avoided transmitting HIV to their babies. The program was very successful. I met its founders when they won a Skoll Award to honor the impact of their direct programs. However, after years of trying to scale up in different countries, they shifted to working indirectly by sharing their innovation with health ministries for adoption. As a result, the reach of their innovation far exceeded the impact they could have had directly.

Your challenge, whether you are a long-standing incumbent or a new innovator, is to look well beyond the status quo. The people and communities you serve deserve better. New technology can make completely new programmatic approaches possible, approaches that can be much, much better than current practices. To "kill a dinosaur" is to be bold. Tech is not cheap. It doesn't make sense to invest a large amount into an incremental tech solution that makes things only 5 percent better when a similar-sized new approach has the potential to make things five times better!

**Exiting Arkenstone, Starting Bookshare: How to Sell a Charity**

I personally dealt with killing our own dinosaur in the case of Arkenstone. An investor wanted to buy it, but I was resistant. However, our organization lacked the funding to start any new social enterprises. The chance to fund new innovations ended up overcoming my reservations. We received roughly $5 million from the investor that purchased our social enterprise from us. It was a complicated process, but we were able to convince the California Attorney General's Office that the price was fair and that the resulting funds would be used to start multiple nonprofit social enterprises (which is what happened). The Arkenstone products continued under the new for-profit owner. Amazingly enough, the flagship Arkenstone software application, Open Book (which replaced the Arkenstone Reader when PCs became powerful enough), is still available in 2025, more than twenty years after we sold the enterprise. However, as a technology it is a dinosaur whose time has long passed.

The funding from the sale of Arkenstone came at just the right time. When we were negotiating the sale, I had my Napster music session with my son Jimmy (described earlier in this chapter), and I had a new idea just waiting for money to invest in trying it out: Bookshare!

Within two years of its founding, Bookshare had more than fifteen thousand titles available for download, heavily weighted to the titles people wanted most. Increasingly, people who had print-related disabilities but didn't own a scanner were able to benefit from those who had one. The lightening of the financial burden was significant: it cost more than $1,000 to turn a PC into a reading machine but only $50 a year to subscribe to Bookshare. In addition, Bookshare worked on a mobile phone as well as a PC.

At the same time that we were reinventing ourselves by exiting the Arkenstone reading products and moving on to become an online library, we were also taking on the incumbent libraries for the blind and people with print disabilities, which were much older dinosaurs than Arkenstone! Our main competition at the time was Recording for the Blind & Dyslexic (now renamed Learning Ally). Putting books on tape for veterans and children who were blind or had a vision impairment was a huge innovation in the 1950s. However, when Bookshare came along and started disrupting the field in the early 2000s, we realized that many of the nice people working for and volunteering for the talking-book libraries thought their mission was to have sighted people help blind people. They saw the idea of removing the need for sighted people to read books aloud as being against their mission.

However, as a new entrant, the Bookshare team saw the mission as not being about a particular method of charitable action but instead about solving the problem in a better way by empowering the blind community to help each other. It turned out the blind helping the blind was incredibly effective. When I first came up with the idea of dealing with dinosaurs, I was thinking about these highly traditional organizations in the field of providing services to blind people.

Compared to the status quo in the field of libraries for people with print-related disabilities, Bookshare delivered an accessible book digitally for less than one-fiftieth of the cost of sending an audiocassette tape through the postal service (which went on for many years even after the cassette tape became obsolete in mainstream society). That's just one example of the kind of leverage new technology can offer: it is often possible to imagine a new approach that offers a ten-to-one (or more) reduction in cost to deliver a specific item of social good, such as an accessible book to a person with a disability. The advent of Bookshare also put pressure on the incumbent libraries to improve their programs, lower their costs, and modernize their technology.

Other examples of tech-based social innovators who took on the dinosaurs of their fields (sometimes dinosaurs they had helped create) include:

**Community Solutions.** Community Solutions was fighting homelessness in the traditional way with traditional metrics by running a successful campaign to build one hundred thousand units of housing. And yet homelessness remained a persistent problem in the communities in which Community Solutions' staff had been working. So they flipped the metric with new technology. Instead of measuring success by how many shelter beds were occupied (or built), they kept track of everyone in a community who was homeless as of three months earlier and tried to ensure that all those people were housed. It takes boldness to abandon a widely accepted strategy (building beds) and shifting to new way of looking at the problem. I go deeper into Community Solutions in chapter 8.

**Crisis Text Line.** Crisis Text Line was founded when DoSomething, a teen volunteerism nonprofit, started getting texts from teens in crisis after it sent out material on volunteer opportunities. The founders recognized that traditional toll-free suicide prevention hotlines were not seen as an option by teens steeped in texting and social media, so they set up a pioneering helpline that was 100 percent texts. Now the Crisis Text Line is a few hundred million texts into helping teens in the United States and a few other countries.

**Medic.** Medic went through a few sizable transitions over time. Tech innovations included evolving from an SMS-centric approach to communicating with health workers to building Android mobile applications. But the biggest transition was the shift to the open-source Community Health Toolkit, where Medic made the effort to share its core technology with many other organizations supporting community health workers, which was highly successful. By giving its core technology asset away to the field for others to adopt (and effectively submerging its brand), Medic ended up having far bigger impact.

**TechSoup Global.** TechSoup Global started life as CompuMentor with the goal of matching up technically skilled volunteers with nonprofits in the early days of the wider adoption of the PC in community-based organizations. It pivoted to become the largest operator of donation programs by the for-profit tech sector by creating a practical solution for scaling up the availability of deeply discounted packaged software applications and computer and networking hardware. Along the way, it built the largest database of nonprofits around the world, facilitating the flow of grant money as well as technology. Most recently, it has made yet another huge change as the software world shifts from standalone software applications to cloud-based solutions. TechSoup Global is a terrific example of an organization that keeps reinventing itself and achieves much larger impact with each iteration.

Social entrepreneurs often start by identifying the dinosaurs in a social sector field that they care about and coming up with a new innovation that significantly improves on the status quo. This is the motivation driving the majority of the most exciting nonprofits started in the twenty-first century, such as those winning the Skoll Award for Social Innovation.

However, being recognized with a big award is not an excuse for sitting on one's laurels and turning into the next dinosaur. The leaders with the most impact over the longest term are often those willing to reinvent their organizations, sometimes more than once, whether with a fundamental tech overhaul or a change in how they measure success or a move to work at the field level.

## GOOD IDEA 3: AIM FOR THE CLOUDS WITH DATA

The final great tech-for-good idea involves data and leveraging cloud infrastructure. Modern business has moved the great majority of its software into the cloud for excellent reasons. The cloud delivers more powerful tools for less money and works reliably on just about any data-enabled device. Solutions that were simply not possible in an earlier tech era have become practical because huge amounts of data can be stored and processed in the cloud, and businesses are able to connect directly with networked sensors to measure all kinds of interesting data. Shared infrastructure and tools based on solid tech standards are the future of the social sector, just as they are already in any modern business.

Even more exciting is the fact that the current AI revolution, from voice recognition assistants to generative AI, has been made possible by cloud computing and the vast quantities of data available there. AI is so important to tech-for-good nonprofits that I've devoted all of chapter 5 to the subject. Shifting into the cloud with your data is a prerequisite to reaping the benefits of AI in the future.

Instead of software residing on your PC (or your smartphone), it is built to operate over the internet: the software is running on computers sitting in a data center somewhere in the world. Your data are also stored in data centers. The software running in the cloud is more powerful and has access to far more data than can be stored on your local PC. The people in your organization can access the same shared information easily. The software works with a wide variety of devices and operating systems: you typically just need a web browser to access cloud software. This approach greatly increases reliability as technologists focus on keeping one cloud version of the software running on commodity cloud processors and data storage. If a big news article hits, and the usage of your software goes up by a factor of ten, your cloud supplier is more than happy to rent you more processors and storage, instantly and seamlessly. You pay only for what you use. The cloud is so much better than the old way of running your own servers in a data center or the even older way of having your own server in a closet in your office, where your capacity was fixed and you were on the hook for any server crashes.

Putting your software and data in the cloud gives your new project the chance of global reach if you want that. When my nonprofit launches a new cloud software platform in English, there are potentially users in more than one hundred countries instantly. We can also easily translate a product into another language without significant expense, which creates leverage for the right kind of project.

Of course, putting valuable data in the cloud puts the data at greater risk of access by unauthorized individuals. However, the calculus is that it is better to have your confidential data in the cloud with strong protections and access controls than not to have the data available for use because they're stranded on one person's PC. And, of course, storing valuable data on a laptop computer is also risky, as many people have learned by leaving their laptop in a taxi or an airport, causing a major privacy breach and loss of data.

As nonprofits move more and more of their activities into the cloud, as they inevitably must do, they need to take security seriously as the price of receiving all of the power and capabilities offered by cloud solutions. However, when you are using a cloud platform operated by someone else (such as Amazon, Google, or Microsoft), you have the benefit of hundreds of expert software engineers working to defend your data—a much better scenario than the old days of data security depending on a single tech person on your team or on an outside consultant.

The great majority of exciting new tech-for-good enterprises are cloud based for all of the good reasons I've just outlined. I will highlight just two, Nexleaf Analytics and Kobo Toolbox.

**Nexleaf Analytics.** Nexleaf Analytics works on assuring the stable temperature of vaccines for more than 14 percent of the world's babies each year. Vaccines need to be kept at a certain temperature range; letting them get too warm or too cold for too long renders them ineffective. Nexleaf's primary hardware product is a temperature sensor that goes into a vaccine refrigerator and regularly transmits the temperature measurements into a cloud software application. The software is where the powerful impact happens, both for individual clinics and their refrigerators as well as for entire vaccination programs. If a refrigerator stops working correctly, the health clinic staff will get an alert on their mobile phones. They can investigate and do something to preserve the vaccines: start a generator if the power is out or move the vaccines to a backup refrigerator or call for a specialist to repair the refrigerator or any combination thereof.

The goal is not to use the software and the data as a club to punish medical staff for not maintaining the correct temperature but as an indispensable tool for doing the job they are committed to doing, which is to deliver critical (and potent) vaccines into the arms of infants. In addition, health ministry leaders at the district, province/state, and national levels get information they can use to better run their vaccination programs. Simply knowing how many vaccinations were administered or how many vaccines were spoiled is not enough. More detailed data help these leaders understand if their clinics are having problems with reliable power, spare parts for equipment, the transportation system, or staff training, so they can take action to improve their programs.

**Kobo Toolbox: Surveying the World**

Kobo Toolbox is designed to make it very easy to collect data using mobile phones, even in places where there is no internet. The Harvard public health professors Phuong Pham and Patrick Vinck founded it in 2005 with the goal of replacing data collection on paper forms and the manual transfer of information on paper forms to digital files with an entirely digital process.

Not surprisingly, the need to collect data is common across the social good sector. Kobo is used to collect public health data, capture baseline community data before and after a natural disaster, track services to refugees, and even document attacks against health-care facilities (as in Syria). The scale of Kobo is immense: more than fourteen thousand organizations (including "virtually all United Nations agencies") used it to collect more than two hundred million survey responses in 2022.[7]

Readers are probably familiar with commercial web-based survey tools such as SurveyMonkey: imagine if someone were to design a survey tool with the entire world in mind. It would need to have a particular focus on the social sector's needs and operate in areas without internet connections by syncing up with the cloud later when a connection is available. It would need to make it possible to collect data from everybody, even people without phones. Oh, and imagine if it were free and open-source software, could host any organization's data on its servers, wouldn't claim ownership in that data, and would offer cloud-hosting services for free to the great majority of users. That's Kobo. It's the real deal!

The Kobo software makes it easy for any organization to create a survey to collect a wide variety of different kinds of quantitative and qualitative data to meet its particular needs. Over time, Kobo has added more and more functionality to make these surveys richer in content by adding the ability to collect location data (that is, GPS coordinates) and to record audio. Once the survey is designed, Kobo then makes collecting the data possible through a variety of channels. Surveys can be completed by respondents with their own devices, or survey takers can visit remote areas and use Kobo to collect data from people without devices or connectivity. Then it is easy to combine the data into maps, reports, and visualizations inside Kobo or to export it to other data tools.

For its first fourteen years, Kobo was a project of the Harvard Humanitarian Initiative, a university-wide research center that works on disaster response and humanitarian challenges. In 2019, Kobo spun out of Harvard to become its own independent nonprofit organization.

With a budget of roughly $1 million per year, Kobo is proof of the kind of leverage that a tech-for-good nonprofit can have. Its founders saw a need for better data collection tools, built an open-source project on top of another open-source project (ODK), provided subsidized cloud data services, and made it far, far easier for tens of thousands of organizations to obtain the data they need to respond to people in need.

Kobo is a great example of all three best tech-for-good ideas in this chapter. Its tools are precisely targeted at equipping the front lines—the survey takers and the community members in search of assistance. It was founded to kill off the paper form as the primary way nonprofit agencies collect information. It makes the cloud work for all of its users by both offering servers to host the data for less technical users and making it easy for organizations to set up and operate Kobo servers.

## CONCLUSION

I have just scratched the surface of the advantages of going with cloud-based solutions. Chapter 8 takes this discussion even further by talking about the ways the cloud can shift power to communities, enable large-scale data collection, and allow AI to be used for social good.

Our three best tech-for-good ideas adapt powerful learnings from the tech industry to the needs of the social sector. These learnings create opportunities for leaders who want to effect dramatic improvements in their organizations and their fields.

By equipping frontline staff or, better yet, community members with effective tech tools, we can work to make humans smarter and more capable. The social sector has found that working with communities instead of "doing programs to them" usually works much better.

By keeping an eye on how to kill the dinosaurs, we will be open to new innovations that break with past traditions and might well be a better way to address economic privation or social injustice. Whether by reinventing our existing nonprofit or creating a new one to do a much better job of delivering value, we should always be open to delivering more social benefits that are more affordable and more welcome in the communities we serve.

In addition, by being mindful of the power and scale inherent in data and cloud-based technology solutions, we are likely to come up with great ideas for creating far more impact and thus begin to take advantage of the power inherent in the large-scale collection of data, but now for the purpose of doing good!

Now that I've discussed the three big picture patterns of how to use technology for social impact, it is time to describe in much more detail how to build a successful tech product.

# 4 DESIGNING TECH FOR GOOD

The ramifications of deploying technology for social impact go well beyond the direct impact of the technology itself. Technology unlocks new possibilities on how to solve problems. It helps us know what is working and what is not, thus allowing us to jettison plans that we were certain would work but do not. To make an impact with technology, we have to change the sector's approach to design.

I previously discussed the social sector time machine, noting that in many nonprofit organizations the technology sitting on the staffers' desks and the software they are using are obsolete. Another aspect of the time machine concerns the approaches for designing both programs and tech-for-good applications. Over the past couple of decades, modern tech companies and the tech they create have shifted to completely new design paradigms, such as human-centered, agile, and lean design as well as rapid prototyping. Businesses have adopted new design and planning techniques because such techniques simply work better. It is past time for nonprofits and government agencies to follow suit and adapt these improvements to the goal of making more social impact instead of more profits.

Before we can begin to adapt these approaches, however, we need to understand how they took over the for-profit tech industry.

## WATERFALL DESIGN AND FIVE-YEAR PLANS

When I started in the tech industry many years ago, every major company developed its own software to manage all aspects of its business. Technology projects were designed in a linear way that flowed, now known as the "waterfall model." A project was started, requirements gathered, the overall design made, the software written, and the resulting product tested, launched, and then operated and maintained. Each phase followed the other in a predictable sequential manner, which is where the *waterfall* descriptor originated. It frequently took years from the inception to the launch of a new project.

The waterfall approach is well suited to certain kinds of projects. For example, it makes sense to pursue a linear approach when you are building a house, a ship, or a dam. However, the waterfall approach turned out to be a disaster for innovative products and the goal of creating something new. New products used to be designed and built in secret, often going way over budget, and then launched with great fanfare. And then, more often than not, the anticipated customers or users hated the product, and it failed.

Similarly, companies carefully composed multiyear strategic business plans in which they confidently predicted what they would be doing in the fourth quarter of the third year of the plan. People had a hard time imagining how to run an organization professionally without having a detailed business plan acting as a roadmap for the future. However, the world usually does not cooperate with these detailed plans. Economies go into recession, new competitors disrupt the market, and inflation challenges the business model. Most of all, the pace of technological change keeps increasing, pushing most of business and society to evolve as a result.

It's no surprise that the innovation end of the tech industry eventually became disenchanted with the waterfall model and long-range planning. Too many products failed, too many billions of dollars of venture capital were wasted, and too many companies struggled. There had to be a better way.

## SILICON VALLEY BECOMES MORE AGILE

Luckily for the tech industry, new design approaches were developed to do a better job of creating innovative products. Human-centered design and design thinking were pioneered by faculty at Stanford University more than sixty years ago and hit their stride in the 1990s and 2000s. The idea was for you (the designer) to engage with prospective users of a new product early in the design process. For a consumer device such as a phone, you might show someone an early physical mock-up (often just a piece of foam) and ask them to imagine using such a device and putting it in their pocket. This is how many early tech devices were initially designed before people put them in their pockets. You could build a series of screen images that showed how a software product might look and what it might do.

As the design progressed, you (as the designer) could keep circling back to users and getting their feedback, making the product incrementally better. By the time the product was ready to launch commercially, you would have a pretty good idea that it would be welcomed into the market. This process helped you avoid spending years creating a product that nobody would buy. In addition, if it turned out that nobody liked your initial product concept, you could stop the project early, having spent only 5–10 percent of your planned design budget. Or you could pivot to a completely different prototype product based on what you learned from users about what they really could use.

This approach to design was eventually adapted to the software industry, with approaches that had names such as "agile software development" and the "lean startup." Entrepreneurs (and venture investors) learned that the biggest question for any new startup concerned "product-market fit." In short, will users actually use your software? In a lesson rich with resonance for the social sector, designers became far more interested in what human beings *will* do with technology rather than what they *should* do. Making money depended on user behavior in the real world, not on how we wish users would or should behave in a theoretical world.

A common maxim among venture capitalists is that they would rather invest in painkillers than in vitamins. Far better to have a product that addresses a real user need, relieving a source of struggle and aggravation, than to make a product that is merely "nice to have." Vitamins may be good for the user, but the benefit is harder to detect. Many of the applications unlikely to work in real life (like those in chapter 2, "The Top Bad Ideas in Tech for Good") are more like vitamins than painkillers.

So, how can you learn about user behavior? The good news is that modern cloud-based solutions make it very easy to collect data: cloud platforms throw off tons of data as part of their intrinsic nature. It becomes very easy to see (by collecting usage data) how much time prospective users will spend with your solution or if they will come back to your product a second time. It's easy to see what users are doing with your product even if you don't look at their confidential information. Figuring out if you have a successful product in the making becomes easier. If users quit using your product after two minutes and never come back, you have a dud on your hands. Time to go back to the drawing board.

However, assessing the performance of software prototypes doesn't stop at your current product vision. It is also easy to find out why prospective users are checking out your prototype and to learn what kind of challenges they actually have. From these insights comes the idea of a *pivot.* If the users don't like your initial product but are willing to share what they really want, you should listen carefully. Perhaps they are very excited about a secondary feature of your product but not its main feature. If three users mention a challenge unrelated to your product that aggravated them enough to bring it up, you might want to check to see if there are many more users with the same challenge.

The goal of early user testing is not to force your initial idea on your prospective users but instead to learn from them. By listening to these users, you will get insights into their deeply felt needs. Smart innovators pivot as they learn. They set aside their original ideas in favor of much better ideas based on listening to people, encountering reality, and recognizing opportunity.

The user-centered revolution in software design sparked a move away from traditional software design specifications. Product managers now collect stories from users that focus on the benefit to the user and are in accessible language. For example, "I would like to have a report that summarizes the top three issues facing our employees." It's pretty easy to build a quick mock-up of such a report and ask users: "If we built a report that looked like this, would it meet your need?" If so, then the software developer can build the actual software to transform a mock-up into a real report generator. The job of the development team shifts from building a product to a design specification to making the user's wishful story become true.

My first successful for-profit and nonprofit products were started in the era of old-fashioned approaches such as waterfall design. Our projects came in late and over budget every time. I thought this was just how things were. The process changed for me only when my nonprofit Benetech started working on its first product for the human rights movement Martus. Our goal was to build a software tool for human rights defenders to collect the stories of survivors and witnesses to abuses. Our original plan was to use the standard design approach we knew, which was the waterfall. Kevin Smith, the senior developer who led the writing of the software, came to me and made a case for changing our design approach to the new, more agile techniques. I reluctantly agreed to give them a try. Kevin mocked up how Martus would work as a series of web pages that looked like a working product. The demo was so convincing that more than one prospective user thought Martus was actually working software, and they wanted it *now*. We had to explain that it wasn't done yet, but the success gave our team the confidence we were on the right track and to keep going.

We went around the world getting feedback and built the product based on input from human rights defenders. Many improvements needed to be made to the initial mock-up, but Kevin and his team were able to deliver the first version of Martus pretty much on time and on budget. That had never happened to me before. Even better, human rights defenders started using it because it had been built to meet their needs and the needs of other people like them. I joined Kevin as a believer in these new design approaches.

## WICKED PROBLEMS

The old-fashioned design approaches worked reasonably well when the problems were straightforward. However, they struggled with more complicated situations. I have highlighted the benefits of new design approaches when trying to make software that humans will regularly use happily. But there's more. More agile approaches, especially when combined with much more data collection, enable us to be more ambitious and tackle much bigger challenges—for example, *wicked problems*.

Many social problems are wicked problems: they are tough to describe, their cause(s) and effect(s) are not obvious, and traditional problem-solving approaches don't work well enough with them. I am not labeling them wicked problems because they are evil (although some are), simply using a known term in the planning and policy field. There is a reason why challenges such as poverty, domestic violence, and the climate crisis persist: the behavior of billions of humans interacting with economic systems, governments, and culture is complex. If these challenges were as easy to solve as a disease that can be cured with cheap single-dose drug, they wouldn't be wicked.

If you've worked in social change, you've seen the impacts of wicked problems when unintended consequences arise. You may make progress on a sustainable development goal, but a by-product of that progress might be that you make something else worse. What if you made land occupied by the poorest of the poor more valuable, hoping to help them, but instead more powerful people steal that land (which has in fact happened)? So many social problems stem from the behavior of multitudes of humans. You may not understand the incentives and disincentives influencing behavior in areas such as social and cultural norms, economic pressure, health issues, and a host of other challenges.

The recognition of the connections among different problems and different sectors has been shifting the behavior of the social sector. For example, a large part of the conservation sector has shifted from solely criticizing the business sector to engaging with industry to shift incentives for removing toxic materials from their products or building fewer coal-powered plants to generate electricity. The agroecology movement recognized that it was going to be impractical to advance biodiversity without engaging farmers because of the outsize impact of agriculture on the typical landscape. (My simple definition of a landscape is an "ecosystem plus humans.")

Technology makes it possible to tackle wicked problems. It allows you to ask good questions about what is happening in the real world. And, based on what you find out in the responses, you can come up with even better questions. When I started a humanitarian landmine detector project, I thought that saving time would be the valuable innovation. But that was before I found out that demining is frequently initiated as a jobs program for demobilized rebels and soldiers following a peace deal to settle a civil war and that reducing the number of deminers needed was not helpful! You might learn that the real reason why few girls are going to secondary school in a particular place is different than what you believed at the start of your project. This new information allows you to pivot your efforts to something that is more likely to realize your mission goal of increased female participation in secondary education.

Cheap data collection enables rapid learning. Today, it is very easy to design a survey for thousands of people. That being said, more and more people are getting tired of being poked and prodded and questioned all of the time. They get less tired if they see that you are actually listening to them and modify your approach as a result. That's one of the reasons I love agile software development, where the team releases a new version of the product every week, month, or quarter. Users are much more willing to put up with defects in your current product and give feedback if they see steady and regular improvements.

One of the magical things about many tech products is that you can collect huge amounts of data from the behavior of your users as they use (or don't use) your products. You don't always need a survey to gain insights about user behavior. After all, Google and Facebook build their knowledge of their users almost exclusively through observation rather than directly asking users to answer questions, and their discovery of something surprising is particularly valuable to them.

These ideas are not entirely foreign to social change practitioners. The word *holistic* is frequently used to describe the complex nature of a social problem. Tech helps here, too: in place of a tight focus on a single issue, tech allows data collection on a much wider range of issues, enabling a more holistic view of the community being served.

Funders and policymakers are wising up to the problems of narrow thinking. For example, those who are working on the world's food challenges are thinking in terms of regenerative approaches, with healthy food produced by farmers who have a good quality of life and which leave the soil better off, protect water quality, and protect biodiversity. This involves the awareness of a much larger set of problems that requires a more complex solution than simply increasing agricultural yields per acre or hectare of land, the single-minded approach that has been the leading driver of environmental and land degradation.

These new approaches also align with new models of social programs. Rather than through a singular focus on delivering a single benefit, which might not help all the target community, social change can be approached as a partnership between the social enterprise and the people it exists to serve. More and more social sector actors are shifting to supporting the agency of individuals in need because it turns out that this approach usually works better than the top-down approach of doing social programs "at" or "for" people. The right tech solution can be very helpful in unlocking the power of partnership in social change.

## THE STRATEGIC PLAN IS DEAD

Notwithstanding the fact that many foundations I know are in an endless series of strategic planning exercises (smile), the traditional five-year strategic plan should be dying a well-deserved death in the social sector. One of my favorite articles on this topic is "The Strategic Plan Is Dead. Long Live Strategy" in the *Stanford Social Innovation Review*, and it dates back more than a decade![1]

What did organizations with a fixed strategic plan do when the pandemic hit in 2020? Most of them had to throw it out the window as they pivoted to online work for their staff and online programming for their target communities. Even medical organizations delivering in-person health care had to make massive changes to how they delivered that care. Rigid organizations that kept to their old plans didn't do very well.

In today's unpredictable world, the rigid strategic plan should be replaced by a more flexible and agile approach, grounded in the organization's mission. Focus should be on the "what" of a social goal (mortality down, good jobs up, sexual assaults down), not on the "how." How will we know if we are accomplishing the "what"? What particular value do we bring to solving the goal? What other capabilities do we need to make an impact?

With this kind of strategy framework, if a new opportunity arises that does a better job of advancing the desired social goal, the agile organization can respond to that opportunity rather than resisting it. If your entire organization is clear on your social good objectives, and your team is open to sensible and ethical experimentation with how you deliver these objectives, all of you will be empowered to find and test better programs.

An interesting example of this issue in my experience was Recording for the Blind & Dyslexic (RFB&D, which has since changed its name to Learning Ally). RFB&D was the largest audiobook library for students with reading-related disabilities, such as vision impairment or dyslexia. RFB&D acquired a small nonprofit founded by George Kerscher, Computerized Books for the Blind, which was pioneering the idea of accessible e-books. In place of tapes delivered by mail, e-books could be downloaded and converted into voice, enlarged text, or braille. However, not long after acquiring Kerscher's fledgling nonprofit, RFB&D shut down the e-book project, preferring to keep its focus on having volunteers narrate audiobooks for people with disabilities.

This decision created the opportunity for Bookshare to come along years later, borrow Kerscher's brilliant e-book idea (with his blessing), and eventually take the federal contract for providing accessible books nationally away from RFB&D (a story I recount in more detail in chapter 7). My perspective on this was that RFB&D in those days thought its mission was to have sighted people read to people with disabilities—a focus on the "how" to deliver a social benefit. This vision created an opening for Bookshare as a new entrant to deliver the "what" of accessible books with a different "how" (digital text e-books rather than human-narrated audiobooks).

If you are building technology to support your mission, it makes sense to take a less rigid approach to planning. The same kind of thinking that increases the likelihood of new tech product success will usually yield higher success in the design of entire programs. Rather than building fixed assumptions into your program design, you can build data collection into your project to test those assumptions early on in the project. If one of your key assumptions is wrong, you can quickly adjust your program design rather than staying the course.

It's perfectly appropriate to be doing cutting-edge programmatic design while using technology that is nowhere close to cutting-edge. In most tech-for-good applications, the hard part almost always involves the humans and their behavior, not the tech! Build the technology to give you the information you need to improve your social outcomes, and your program planning efforts and eventual positive impact are much more likely to follow.

## PEOPLE-CENTERED DESIGN FOR SOCIAL GOOD

The social change field has something that Silicon Valley doesn't have: a strong commitment to social justice. This commitment gives the tech-for-good team a moral compass. We should be trying to achieve product-user fit (more than product-market fit since the market is failing to meet the need we see). At the same time, we should be advancing the values shared by the great majority of the social good sector. This goal is more complicated than the simpler calculus in the for-profit sector, which is primarily whether the business is succeeding financially. Whether your social enterprise's customers are individuals or organizations, how you operate will be scrutinized. To make a positive social impact, your actions need to generate trust. As I noted in chapter 1, I believe trust, not money, is the currency of the social sector. Being trustworthy gives you the license to operate!

Most of the following three tech-for-good design maxims—empowering your users, achieving greater accessibility, and increasing team diversity—should also be how the for-profit world works. Certainly, some companies have committed to most of these principles. However, they typically see them as optional, not required.

User empowerment is a core value of social entrepreneurship in general and of nonprofit tech-for-good efforts in particular. Accessibility should be considered in both of its aspects: easily used by people with disabilities and reaching the widest possible audiences (rather than focusing solely on those with more money). Team diversity is even more important in tech for good because a diverse team is more likely to be able to put themselves in the shoes of the communities to be served.

In the real world of limited funding for social change, it will be impossible to follow all three of these design imperatives all of the time. However, you should be thinking about them all and figuring out which ones are a requirement for a viable social enterprise and which ones might be progressively realized over time as your product succeeds and expands its reach.

### Empowering Your Users

First, you should be empowering instead of disempowering your users. Authentic social change movements are trying to reverse entrenched patterns of injustice, and you should think of your product as an instrument for shifting power. As you are designing your enterprise, are you engaging with women, minority and Indigenous communities, people with disabilities? You should be trying to place yourself in the shoes of your users, not just solving their technical problem but helping to overcome existing patterns that harm or discriminate. This is particularly critical when we're talking about collecting data from vulnerable people about what makes them vulnerable, which is very common in social change. At a minimum, we should be securely storing the data of vulnerable people, as securely (or more so) than we would want our own confidential data kept. Data breaches or oversharing can be a source of disempowerment as data can often be used against people.

How can you ensure that you're not using data in ways that reinforce existing power imbalances? As my current nonprofit, Tech Matters, started exploring the need for better software for local leaders to respond to climate change, our very first interview was with Kamau Mbogo, a Kenyan leader who patiently explained the idea of data colonialism to me. He noted that although he led the regional effort to revive the Lake Naivasha area, one of the best-studied ecosystems in the Rift Valley, he and other local leaders had no access to the data about the place where they lived and where they made most of the decisions. The data about their place was at European universities or in agricultural supply chain companies. Perhaps an international NGO ran a project in the region, and then the data disappeared when the money for the project ran out. Scientific papers about Lake Naivasha cost far too much for the typical Kenyan to download a copy to read.

His observations made me realize that not only were major, mainly American, tech companies extracting data and value from countries all over the world, but nonprofits were also engaging in these extractive practices. The input from Kamau Mbogo and the dozens of interviewees that followed him were crucial in shaping what turned into our Terraso social enterprise, from both a product needs and a behavior (avoid data colonialism) standpoint. As you build a tech solution focused on a specific need, you also need to keep in mind these broader issues to ensure you are truly supporting the people you intend to serve.

### Achieving Greater Accessibility

Second, you should be thinking about accessibility in the broadest sense. My long-term passion is for supporting people with disabilities, who are often the poorest of the poor around the world. A disaster-response tool built without thinking of people with disabilities will miss a huge part of the problem: people with disabilities are disproportionately affected by disaster.

You will not be alone in building products that work with the full range of human capabilities: smartphones and laptops are now built to support accessibility. One of your design objectives should be to be aware of these accessibility features and make sure you don't break them in your product design. For example, tech tools make it easy to take some text and create an image of that text. Unfortunately, when text is delivered to a blind user as a picture, the text is inaccessible to the built-in screen reader on the user's device. Instead of reading the words aloud, all that person hears is "image." So don't deliver text content as pictures; deliver it as text, just like Google or Apple intended!

Further, accessible design is often just simple good design. Many of the features you build in to support people with disabilities will help people with other needs as well. This is commonly known as the curb-cut effect, so-called because the curb cuts required on sidewalks in many countries were put there to help people using wheelchairs but are far more frequently used by people with baby strollers or dragging wheeled luggage.

Since the people who most need technology are the least able to afford it, how will your enterprise reach the poorest of the poor? Financial accessibility questions extend far beyond just the basic price of your product or service. Will the person in need have access to the right kind of mobile device that supports your product? Is there digital connectivity where your users reside? Even if they have the right device and connectivity, will they be able to use your product based on their tech skillset? Even if your product is made available for free to most users, these issues are still worth examining before you distribute it.

Accessibility also includes support for the languages spoken by your community. Almost all nations today have populations who speak different languages. The for-profit product can often get away with supporting only the dominant or historically colonial national language (English, Spanish, French, Portuguese, and so on), but this will often not be enough for you to reach your nonprofit social impact goals. The lack of support for languages by technology products is one more way that minority communities find themselves second-class citizens in their country of origin or as immigrants in a different country. More and more devices are supporting translation as a built-in function, but not for all the languages people speak.

A related language accessibility issue is literacy. Your target community may speak a given language, but what if they are unable to read or write that language? Many tech-for-good projects have chosen to make audio or video a key part of their design, replacing or augmenting digital text. For example, Amplio supports rural farmers in Africa by providing them with audio content, supplementing the limited number of government agricultural extension workers who rarely get to meet with the typical farmer. Many of the accessibility features originally created for people who can't see are also useful for people who can see but cannot read written text in a language that they understand when it's spoken aloud, thus helping people with invisible disabilities such as dyslexia and those who never learned how to read. One big accessibility maxim: make it possible for the same information to be delivered in more than one way.

### Increasing Team Diversity

Third, you should strive for diversity on your product and program development design teams. The tech industry (at least in the United States) is not particularly diverse. A team made up of young men drawn from the dominant culture in one country are unlikely to do as good a job as a team that includes women and people who are from different cultures, speak different languages, and especially have lived experience with the social problem(s) being addressed. The American civil rights lawyer Bryan Stevenson calls on those of us who serve the greater good to embrace proximity.[2] It is difficult to help create a solution to a social problem you have only read or watched a documentary about.

If your goal is to focus on the needs of women, you are likely to be in deep trouble if you build a tech team composed only of men or led only by people who don't value the input of people who are not like them. The mainstream tech industry does this to its detriment, but such an approach does not work in creating social change. A large part of the movement to hold the tech industry to account is focused on the need to diversify who builds the technology we need. If you are not familiar with the public interest technology movement or data feminism, this is a great time to come up to speed to avoid these challenges.[3]

Importantly, the needs for empowerment, accessibility, and diversity don't cease once the product is designed. They should be built into the day-to-day operation of your nonprofit.

**TalkingPoints: Tackling Educational Achievement in a Novel Way**

Heejae Lim founded TalkingPoints, an innovative social enterprise that exemplifies the values of empowerment, accessibility, and diversity. The goal of TalkingPoints is to improve educational outcomes for children, especially those from diverse backgrounds.

TalkingPoints's breakthrough idea was to improve educational outcomes not by building traditional educational software for use in the classroom or at home but to use technology to increase parental engagement. Lim knew that research shows that student achievement is linked to parental involvement. For immigrant parents who do not speak the same language as teachers and school staff, such involvement is hard.

Starting with a website solution, TalkingPoints automated the translation of messages between parents and teachers so that each could communicate in their preferred language while discussing their common priority—the student. By simply empowering teachers and parents to connect, TalkingPoints was able to improve student educational outcomes.

As time went on, and TalkingPoints grew, it expanded into developing a more capable mobile app and making it possible for other school staff to engage with parents. In my podcast interview of Lim in 2024, she noted that TalkingPoints is supporting more than four million families each year, communicating in more than 145 languages.[4] Very few for-profit educational technology companies would invest in supporting that many languages, but as a nonprofit TalkingPoints aims for the widest accessibility to maximize its social impact. TalkingPoints also prioritizes a diverse team, given its mission and the fact that its founder herself has the lived experience of growing up as an immigrant.

I also particularly like TalkingPoints because it is a shining example of success using technologies that so many others try to use and fall short. It has built a highly successful mobile app that is very well matched to the problem it is solving. Plus, it has a long track record of using AI well. TalkingPoints was originally developed using Google Translate, but as time went on, the team used their unique knowledge about educational topics to improve translations for the app's users. And more recently TalkingPoints has been using AI to help its users communicate more effectively. For example, a message written at a fifth-grade comprehension level is both easier to translate and more likely to be understood.

Not only does TalkingPoints improve student achievement as measured by standardized testing, but it also has had a positive financial impact on school districts that use it. Student attendance has increased, which translates to increased funding from the state education authority amounting to more than $150 a student, which exceeds the per student cost of TalkingPoints by more than a factor of ten. Not surprisingly, this is a positive development for the financial sustainability of TalkingPoints. The product is free to parents and educators, but more and more school districts are paying to get enhanced features.

I had the chance to see TalkingPoints as it was first developing. Lim was a Stanford MBA student, and I was a volunteer judge at Stanford's annual social good business plan competition. She had already been engaging with the Oakland School District in California, one of the most diverse school districts in the country, and had a solid vision of the fit between the need and her proposed solution. In contrast to my usual response to student business plans, I remember thinking when I heard her presentation, "Wow, this could really be successful!"

### VALUE OVER THE LONG TERM

Social problems are not solved overnight. To have a major impact, organizations must dig in deeply and learn and adapt over time. The most successful organizations doing social good have typically been working away for one or more decades. To be a reliable partner in social change, you need to commit to the long haul.

The same should be true of tech for good but often isn't. The norm in technology work in the nonprofit sector is short-term projects. This is the same kind of thinking that contributes to making some of the poor choices around technology previously discussed in this book (see chapter 2). The belief that you just build the software then you will be done is wishful thinking. This short-term view doesn't work very well in a modern world of cloud computing and rapid learning.

Unfortunately, the short-term view is encouraged by the current state of funding for technology inside nonprofits. When tech spending is seen as something separate from spending for program activities, it ends up in a sort of nonprofit no-man's land, either as overhead or as the funding of one-off separate projects. It is worth exploring the disadvantages of these two treatments.

If tech investments are classified as overhead, they are effectively discouraged. Many donors regard overhead as an evil tax on the real social good work. Years of campaigning against the idea that low overhead rates are a sign of effective nonprofit management have only partially worked. For example, the Ford Foundation used to cap overhead rates at 10 percent and changed that to 20 percent in 2016, but few other foundations have followed suit. Many donors limit overhead to 15 percent or less: I have seen donors with 7 percent and 10 percent limits and saw a recent grant application where the donor rather proudly refused to spend anything on overhead. Doesn't that send a message? Don't invest in tech, don't invest in your people, don't pay for the true cost for management, fundraising, and accounting! It is also a mighty awkward message to receive as a nonprofit founder because the founder's salary (or, at minimum, the portion spent on fundraising work) is often classified as overhead. When I used to start regular for-profit tech companies, these kinds of restrictions would have been unimaginable.

Treating a tech project as a one-time expense has different disadvantages. The global development world has long and bitter experience with the myth of one-time investments, their negative outcomes made transparent by empty school buildings with no staff and wells abandoned because they were not maintained. Nonprofit tech used to work the same way: find a one-time grant to pay for a phone system or laptops for all the staff or for the creation of a CRM system for tracking clients. Short-term grants were made on the assumption that you were supposed to know exactly what you needed up front. This led organizations to commit to the wrong tech, which they might be stuck with for years and years as it slowly (or quickly) decayed.

Even good solutions have problems that pile up over time: the phone system needs to be upgraded to one that handles texts; the laptops break (or become obsolete); or the CRM system goes down (or diverges from what the organization needs). Turnover means that the people who used to (sort of) know how to use the tech tools leave and are replaced with people who lack access to organized training. Now, as more and more technology shifts into the cloud to gain access to the power of data and modern tech capabilities, cloud solutions are never "one and done." Cloud solutions trade up-front costs for relatively low monthly costs and also need constant attention to maintenance, especially around security risks, because the benefits of cloud solutions come with the challenges of being on the internet. Cloud solutions cause the approach to tech as a one-off investment to break down completely.

Unfortunately, most donors are used to funding tech (when they do fund it!) the old-fashioned way, which pretends that you can buy technology solutions with a single up-front purchase without ongoing expenses.

Over time, modern businesses and sophisticated nonprofit organizations have come to realize that the long-term cost of adopting a given technology solution is much more than the initial purchase price or implementation cost. The situation is similar to buying a house: over time the cost of the mortgage, insurance, and maintenance adds up. In the tech field, this is called "total cost of ownership" (TCO), where one tries to identify all of the costs associated with a tech implementation decision over the full number of years it is expected to be in use.

The true costs to a nonprofit committing to new technology for a period of years always far exceed the up-front costs. TCO includes the up-front costs to buy any needed hardware, pay any one-time licensing fees, and pay for the consulting to customize the solution to your needs. But the ongoing operational costs—such as cloud expenses (for storage, processing, and bandwidth), ongoing license fees, maintenance, and training—also need to be factored into TCO. How much of your staff time will be taken away from their regular jobs to become proficient in the new technology? Will you need someone on your team to be the internal point person on the tool(s)? How much of their time will be needed? Assuming you come to depend on the new technology, there will come a day when you need to replace it when it becomes obsolete. None of this is optional or unlikely and thus needs to be considered in advance when making a major tech solution.

I remember being asked by a number of nonprofit CEOs over the years about the CRM software from Salesforce. "I hear Salesforce is free" would usually be one of their first comments. "Actually, I bet Salesforce will cost you at least half a million dollars, maybe a million, over the next five years," I would reply. Yes, at the time Salesforce would provide ten "seats" of their basic product for free, but the nonprofit might have twenty-five people who needed access, or they might need an additional product that was not free, only discounted. You might end up paying Salesforce $10,000 to $30,000 per year, which is still much less than a for-profit would pay (we assume). The nonprofit often needed to spend $10,000 to $50,000 to have Salesforce adapted to the nonprofit's application requirements. However, the real kicker, the much bigger cost component, is staff time at the nonprofit. The great majority of nonprofits using Salesforce need a half-time or even a full-time person to oversee it. In the United States, the loaded cost of a person with those skills is more than $100,000 per year (unless you get very lucky).

That's how you can easily find yourself with a TCO approaching $500,000 or more over the first five years of using a complex product such as a CRM system. This is not a problem, however, as long as the benefits to your organization exceed the cost over that time period.

The issue of long-term commitment is magnified when moving from organizations using external tech solutions to enable their teams to organizations creating tech solutions with internal tech developers. Software developers are expensive. If you are creating a solution for empowering your internal team, you are responsible for the entire project, including the development, operation, and maintenance of the tech for years, as well as training and retraining your team. Moreover, you will need other professionals to interface between the tech developers and your frontline staff, supervisors, and, if applicable, community members interacting with your technology. All these costs need to be part of your TCO projection and weighed against the benefits.

You might think that the tech development cost will be almost all upfront and much lower in the years after the first version is released. The story rarely plays out that way in real life.

I'm not trying to talk you out of doing technology. Otherwise, I wouldn't have written this book! I strongly believe that tech investments can have amazing positive impacts that far exceed the cost of creating and operating them. I also believe the nonprofit sector can build amazing technology for much less than the for-profit sector would spend, assuming that the nonprofit sector's needs offer a sufficiently large market. However, it is very rare for a new technology to come in, make a major positive impact, and cost very little after the first year.

Societal Thinking (formerly known as Societal Platform) is an Indian group that sees technology as essential to large-scale change. Its staff think about massive change at the very beginning of a project, imagining how they are going to reach tens or hundreds of millions of people within just a few years. The story of DIKSHA (from the word *diksha*, the giving of a mantra or an initiation by the guru) reaching more than one hundred million students in India in less than five years exemplifies the power of this approach. The DIKSHA program inserted QR codes into the standard printed textbooks provided in schools so that teachers, parents, and students could scan the code for additional materials to help them teach and learn the content in those textbooks more effectively. Of course, such scale requires innovative tech design to match.[5]

If you want to develop a significant tech-for-good project, you should be thinking of committing to it for at least ten years. If the tech is important enough to your mission, whether as an organization enabling its team with external tech products, building technology for itself, or building tech products for others, you should be in it for the long haul. The reality of TCO and the reality of the technology of today make short-term thinking impractical.

## CONCLUSION

Designing successful tech-for-good solutions requires more agile and adaptive methods. Software and, increasingly, hardware products are data collection engines. Modern approaches to design emphasize using data to continually improve both programs and technology. Everyone who cares about social change must invest in data literacy and actively use learning as the central organizing principle of their work.

However, tech-for-good organizations blend these approaches with the values of social impact. More than the typical for-profit, nonprofits should be emphasizing user empowerment, accessibility, and diversity as they create and scale their products and services. At the same time, they can't part ways with financial reality. Tech is expensive and needs to deliver outsize benefits in terms of cost savings and impact to justify that level of investment over five or more years. Further, everyone involved should prepare to sustain the commitment to a successful tech product over a period of ten or more years to have a hope of making long-term positive change.

# 5 BUILDING SUCCESSFUL AI SOLUTIONS

Artificial intelligence (AI) is technology that is hard to ignore because of both its promise and its potential for harm. Is it a panacea that will have an immense positive social impact? Or will it destroy the economic viability of disadvantaged communities by taking away much needed jobs? The answers are much less dramatic than the questions. AI is neither a miracle pill nor the key to remaking the economy in the next few years. All of the tech-for-good lessons I have written about apply in equal measure to AI applications. AI is a major technological advance that is important to understand and to apply to social good, but only when it makes economic and ethical sense.

You no doubt have questions. I have some but not all of the answers, starting with the following basic questions.

1. What is AI?
2. What are some of AI's limitations?
3. What do I need to know about implementing AI?
4. What should we do when the AI gets it wrong?
5. What process should I follow when creating a new AI application?
6. How should I approach getting the data we need?

As I answer these questions, I'll explore good and bad examples of using AI for positive social impact.

## QUESTION 1: WHAT IS AI?

I am using the popular term *artificial intelligence* despite the fact that computers are still not intelligent. As a matter of fact, computers are still dumb as bricks. AI is definitely artificial: it's a human-created technology that is designed to imitate certain capabilities of humans. The favored term for this technology among data scientists and engineers is *machine learning*, which narrows the ambition to teaching software programs how to recognize patterns based on data. However, the terms *AI* and its popular variation, *generative AI*, have entered into wide use, and I will stick with using them.

The data scientist Cathy O'Neil's book *Weapons of Math Destruction: How Big Data Increases Inequality and Threatens Democracy* (2016) discusses the many ways algorithms are used to disadvantage the disadvantaged.[1] She doesn't like the term *artificial intelligence* because, as she told me, "AI is a meaningless marketing term that is used to make people feel intimidated. It is not an agreed-upon notion by computer scientists, but [marketers] just use it whenever they want to sprinkle magic onto things."[2]

Cathy also likes to demystify algorithms by pointing out that "an algorithm is a method of predicting the future based on patterns of the past." Humans do this every day, she notes, when planning dinner or whether to drive or bike to work. The key to an algorithm's success is in how a successful prediction is defined. Her main point is that "the people in power optimize to their definition of success." In today's tech industry, success is defined mainly by what makes the tech company the most money, not by how it helps the typical user. I return to this issue in chapter 6.

Because the term *AI* has a somewhat magical quality, it is also a moving target. The scientist who coined it in the 1950s, John McCarthy, was reported to have complained, "As soon as it works, no one calls it AI anymore."

When I cofounded my first AI company in 1982 (Calera, where my venture capital board vetoed the reading machine for blind people, as shared in the preface), humans still beat computers at the game of checkers. Optical character recognition by machines was then at the very leading edge of AI. Our breakthrough was to use millions of different examples of characters to train algorithms that could recognize letters, even in fonts they had never seen before. However, it worked only on machine-printed fonts because human handwriting was a difficult challenge and remains so even today. But even that narrow accomplishment, just reading printed text, is still commercially valuable as well as helpful to people with disabilities. Of course, this was the technology at the core of the reading machines provided by my first nonprofit social enterprise, Benetech. Not surprisingly, an AI tool that could read books aloud was very useful to people who have a print-related disability—even if it's not thought of as AI anymore simply because it works pretty well!

We now take for granted that a computer can beat humans at chess and Go, speak to us, and understand voice commands ("Hey Siri/Google/Alexa, play 'Heart of Rock and Roll' by Huey Lewis"). We also are used to these systems making mistakes: using a voice assistant is usually an exercise on training humans on how to manage AI-powered systems that make many mistakes!

Because AI is increasingly venturing into areas that used to be reserved for human beings, we have a tendency to attribute human intelligence and characteristics to the technology. This attribution also creates marketing sizzle for AI companies busy hawking their products and trying to raise large amounts of investment funding.

In 2024, the world has been going crazy over the relatively new part of the AI field called "generative AI." Most AI tools up to this new wave have been focused on making predictions based on past experience. The success of these predictions turns on recognizing patterns from past data. Is that a letter *e* on the page or a letter *c*? Is that a picture of a cat or a dog? Is this borrower likely to default on a new credit card?

Predictive AI is generally easy to assess: Was its prediction correct? This is true even if some predictions need to be confirmed over time, as in the example of a credit card default, which could be checked after a year. Because of this testable nature, it's also straightforward to assess the performance of predictive AI by measuring the percentage of time it was right.

The twist in generative AI is that it is using statistics not to make predictions but to generate new content, such as text, pictures, videos, and even music. "Making something up" is more or less what we are asking generative AI solutions to do, and they are good at making up stuff. However, we should not mistake these outputs for the products of humanlike intelligence.

Computers don't have compassion or empathy even if the words they produce deliver a facsimile of feeling. They don't understand what they are outputting. The current generation of generative AI tools such as ChatGPT and Gemini are incorrectly described as "hallucinating" when they make up untrue facts. ChatGPT and Gemini aren't hallucinating any more than a rock can take LSD: they are probabilistic word generators that spew out words one after the other based loosely on which words are likely to follow what has come before, predicated on massive amounts of input data. Making sense or describing a real event is not required.

Many leaders in the data science field are not overly impressed with generative AI. An influential scientific paper by Emily Bender and colleagues describes large language models, the tech behind tools such as ChatGPT and Gemini, in the following way: "Contrary to how it may seem when we observe its output, [a large language model] is a system for haphazardly stitching together sequences of linguistic forms it has observed in its vast training data, according to probabilistic information about how they combine, but without any reference to meaning: a stochastic parrot."[3]

Having ingested huge quantities of text, these stochastic parrots have a pretty good model of plausible sequences of words but no understanding of what they mean. Another writer has described generative text systems as "spicy autocomplete."[4] I think of these systems as "spellcheckers on steroids"!

As the current hot part of AI, generative AI systems definitely have value. However, they are not plug-ins for humans, where one can replace a human worker with an AI tool, any more than we can replace writers with spellcheckers.

One of the main points in this book is how we must center tech-for-good applications on solving real problems for real people. Imagine if we enumerated all the problems faced by the planet, nonprofits, and disadvantaged communities that technology could possibly play a major role in helping to solve. My estimate is that today's generative AI is likely to be the key technology in addressing perhaps only a small percentage of those problems (although I expect the percentage to go up over the next five years). My optimistic view is that other kinds of AI might be very helpful today in up to 20 percent of the world's biggest challenges. That's a significant portion! But it doesn't mean you should drop everything just to figure out how to use AI in meeting the social challenges you focus on.

## QUESTION 2: WHAT ARE SOME OF AI'S LIMITATIONS?

Keep in mind how extensively AI tools are shaped by very human constraints. The biggest constraint is bias because the data are always biased in some way. The choice of which data are included (assuming they are even available) in training the AI algorithms determines the performance of an AI algorithm in trying to solve a specific problem. Bias is a big issue when the AI was trained on data that is mostly disconnected from your specific problem! Another constraint is the intent of the human designer(s): Who were they, and what was their goal in designing the system? What was their definition of success? Which issues were prioritized and which ignored? And who was exploited when the datasets were created?

In data science, bias simply exists. It's not necessarily a bad thing: it simply reflects that it is impossible to know everything in the world or even about an individual. All AI tools are shaped by the data used to train them. Getting and processing that data are a huge part of the expense of creating new AI technology. The data that can be obtained will be biased in two ways. The first bias is that no dataset has an omniscient view of the entire universe of the phenomenon it is intended to predict. If we trained an AI tool to recognize speech based only on American radio broadcasts, its performance with respect to Mandarin-language radio would be terrible. If we trained a tool to estimate crop yields only on Iowa cornfields, it wouldn't work so well on African maize fields or Indonesian rice fields. The large language models behind tools such as ChatGPT and Gemini were trained on huge but still finite and biased datasets, mainly in English. They will not work as well or at all in other languages. This first kind of bias is not intrinsically evil, but it can have negative impacts when the tool is used on data that are not like the data it was trained upon. This form of bias is a problem for social good organizations because the people they serve are poorly represented in current large AI datasets. Consider that an understatement: most of humanity is not represented in these datasets at all.

The second form of bias is when AI tools pick up negative patterns in society and replicate them. This kind of bias tends to be much less obvious because it simply confirms our bad collective tendencies, of which most of us are unaware. The classic example is predictive policing based on past policing patterns. If minority men were consistently arrested at a far higher rate than majority men even when the underlying rate of offending was similar, a predictive policing tool trained on that data would inevitably recommend policing patterns that would result in the arrest of more minority men. If a recruiting tool is trained on data from a company full of men, it will recommend hiring more men than women. A biased society will lead to a biased AI tool unless extra care is taken to reduce such bias. Sometimes the bias is so pervasive it isn't even possible to repair. When Amazon, a large, rich internet company, developed an internal recruiting tool in an attempt to automate screening of potential employees, the tool so thoroughly discriminated against women that Amazon had to kill the project entirely.[5]

This picking up of societal-level bias shows up in generative AI image tools. Even though the majority of students in medical schools in the United States are women (yet 39 percent of current doctors are women, a legacy of past discrimination), a recent test showed that if a leading AI tool is asked to generate an image of a doctor, it will depict a woman only 7 percent of the time.[6] When it is asked about those who fill high-paying jobs, it depicts white men far more than women or people of color. It does the opposite for low-paying jobs. Generative AI doesn't spit out what society should look like or actually looks like; it spits out what society looks like through a lens of the stereotypes in the data present on the internet.

An AI algorithm doesn't know right or wrong; it's just trying to replicate what it found in its training data. In a similar direction, automated parole systems have been demonstrated to be roughly half as likely to recommend parole for a Black inmate as for a white inmate even though the impact data show that white men are equally likely to reoffend.[7]

The designers of these parole systems disingenuously claimed that race wasn't one of the factors they used in developing their system, but AI systems are great at finding patterns in data. If the US criminal justice system is historically biased against of people of color, AI systems will find patterns that can serve as a proxy for race (such as zip code or first names). Predictive policing and parole systems trained on that data should be expected to discriminate.

The computer scientist Joy Buolamwini discovered while at MIT that facial recognition could not recognize her dark-skinned face.[8] She then did research that showed commercial facial recognition AI systems generally did a terrible job on the faces of brown-skinned women compared to how well they worked on white men. One has to assume that the constraints described earlier were at work: the facial datasets probably didn't contain very many women of color, and the definition of success didn't have a measure of whether the system was disproportionately biased against women with brown skin. It might not have even occurred to the designers of the system to test for this problem because women are a distinct minority in the workforce that is designing AI systems.

When embarrassing problems like this become public, the AI companies sometimes choose to address the identified problem by tacking on an additional fix. However, this usually doesn't fix all of the similar problems present in the source datasets. Other companies sometimes get away with ignoring public criticism if they are still making plenty of profits. In their book *AI Snake Oil*, Arvind Narayanan and Sayash Kapoor argue that many AI products simply cannot work as advertised and, even worse, are actively causing significant harm.[9]

This brings us to another human constraint on AI: Who is designing the system? Negative impacts against people who do not resemble the designers might not be recognized or get much attention, a factor expected to have been at work in the poor performance of facial recognition systems. The tech and AI workforces are not known for being a diverse community, which means they are likely to have significant blind spots when it comes to the impact of their designs on women and other groups.

Next, we have the challenge of what the definition of success is when designing these systems. As Cathy O'Neil points out in *Weapons of Math Destruction*, concepts such as "justice" and "transparency" don't tend to figure in the marketplaces for data, as opposed to business concepts such as "efficiency" and "profits."[10] This focus on making money leads us to many of the worst problems that exist online. My personal observation has been that when faced with a choice between making more money and protecting children online, the big tech corporations have consistently chosen to make more money. If you are considering the adoption of an AI product, you need to keep in mind that it may be optimized for outcomes that might be very different than what you are hoping for.

Another consideration is how the data have been obtained. Active disputes are going on right now between major generative AI companies and the authors and publications whose copyrighted works were used to train those AI tools. It is also important to remember that so many of our AI advances depend on low-paid, often exploited workers around the world who do the work for tech corporations to make the AI better. Some of this work involves viewing disturbing images or videos of violence as part of training content moderation tools to block illegal content (such as child sexual abuse). We owe it to these workers to ensure that their sacrifices don't benefit just rich corporations and people in wealthier countries but also the workers and their communities and regions. Last, some of these technologies consume a great deal of energy to be trained and to operate, which makes them expensive, perhaps putting them beyond the reach of the typically neglected 90 percent of humanity. The high energy use is also likely to have a negative impact on climate change.

Together these human constraints—biased datasets, varying definitions of success, nondiverse tech teams, and the disproportionate impacts of collecting and processing the data—have an overwhelming effect on how an AI system works in practice. Even as the capabilities of AI technology will inevitably march forward over the coming years, it's not clear that these longstanding issues are going to be addressed effectively. However, knowing that the constraints exist will help you to avoid the most common pitfalls and to design for success based on your mission goals.

## QUESTION 3: WHAT DO I NEED TO KNOW ABOUT IMPLEMENTING AI?

Your organization is already making use of AI, of course. AI tools are built into our word processors (autocomplete and spellcheck), our search engines, and our smartphones. For the purposes of this book, AI applications can be divided into two groups: internal productivity tools like those just mentioned and AI in your program activities.

Plenty of tools in the first group are "off-the-shelf," which means that they are commercially available, perhaps with a free version. They can be useful to many nonprofits for improving productivity in more administrative tasks as long as some of their limitations are kept in mind.

ChatGPT, Gemini, and similar tools are in wide use as writing assistants. These tools have been trained on huge amounts of text and so can easily produce *mostly* plausible newsletters, grant proposals, constituent communications, and blog posts. Note the emphasis: these AI outputs tend to sound authoritative and are generally well written. However, far too many people and organizations get into trouble by sending off AI-generated content without careful review. The press regularly covers these embarrassing incidents. Not long after ChatGPT was debuted to the public, an American lawyer faced the possibility of court sanctions because he had turned in a legal brief drafted with ChatGPT's help but in which the robot had invented cases.[11] The cases sounded plausible but were imaginary. The law firm had failed its professional obligations to its client by handing its work over to an unsupervised robot. The judge in the case was not amused.

This is one of the reasons why off-the-shelf generative AI tools need to be supervised in general and especially in high-stakes documents such as legal briefs and funding proposals. It may well be that a dedicated writer will find themselves working 10–20 percent more effectively with the help of these writing tools, but this assumes that the writer is carefully editing the output from the AI or using the AI as a source of ideas for writing a more compelling piece.

Joan Mellea, the cofounder of Tech Matters, my current nonprofit, was an avid user of ChatGPT, and she felt it saved her 20–25 percent of the time she used to spend writing and editing concept notes, proposals, policies, blogs, and articles.[12] She particularly appreciated its ability to compress a 300-word answer into a 250-word answer or suggest a draft policy. She never took and used the output without reviewing it: it was a source for ideas and a new edit. Her main advice was that you should use these tools only about topics where you are already knowledgeable. Otherwise, a big mistake (or three) could easily make it past you.

It's important to keep in mind that most people's jobs involve a significant number of different tasks, only a few of which can be partly assisted by AI tools. In Joan's case, writing and editing was just one piece of her job, so a 25 percent increase in productivity in her writing tasks may have saved her only 5 percent of her total time. When organizations start talking about laying off half of their fundraising team to replace them with generative AI, I get very worried that they will raise much less money. The same goes for software development teams: we and our peers are finding AI tools to be useful in coding, but not revolutionary. A 5–10 percent productivity increase is definitely worth the price of these tools, but they don't change our staffing levels or revolutionize the work. Over time, we will hope for more productivity gains, but the story of software development has already been one of steadily increasing productivity over decades thanks to steady tool improvements.

Another administrative area making effective use of generative AI tools is fundraising (beyond the writing of grant proposals). Since so many nonprofits raise money from individual donors, for-profits are creating tools that make researching donors and customizing messages to them more effective. These kinds of products are also reasonably easy to assess. Do we raise more money without irritating more donors than when humans were writing the messages? Of course, this depends on whether you are comfortable with annoying any donors with AI-generated messaging!

It might surprise you that no generative AI was used in the creation of this book. I made a conscious decision not to use these tools even though I am generally a big fan of AI tools. My goal was to write a book that conveyed my personal point of view about tech-for-good topics rather than producing a bland but well written take on these issues based on what's on the internet (and without any exclamation points!).

I now get cover letters accompanying job applications that are obviously written by generative AI, and though it's good to know that these candidates have the skill to operate a software program, this skill isn't enough for me to hire them.

There is another big drawback in using off-the-shelf generative AI tools: data ethics. Most of these tools are constantly gathering new data to improve themselves, and one of the top sources of new data are the queries fed into the system by users. Many organizations are rightfully freaked out that confidential information from their organization, possibly from their clients, might be fed into a commercial AI system to be stored and used for profit-making purposes. Although the vendors of these tools often claim that the confidential data they have gathered and fed to their AI systems can't be pulled back out of those systems, these claims have been repeatedly disproven. To meet their legal and ethical obligations, more and more organizations are adopting policies against using generative AI tools internally or putting limitations on their use. Even the vendors of these tools have started to limit their ability to use the data of their customers, especially their corporate customers who might simply decline to use a tool that appropriates their confidential content.

Of course, the recent excitement about generative AI tends to overshadow the previous fifty years of progress in AI. A huge variety of AI-powered tools are finding patterns in every kind of data—from images to audio, video, financial numbers, and much more. A good book on this topic (which predates the generative AI craze) is *The Smart Nonprofit: Staying Human-Centered in an Automated World* (2022) by Beth Kanter and Allison Fine.[13] Modern tech-for-good organizations need to keep AI solutions in the back of their mind as they find new challenges in their fields—much better than having AI solutions in the front of their minds, searching for nails for the AI hammer!

There are many other off-the-shelf AI-powered tech tools. Voice recognition products such as Otter are being used in conjunction with online meeting platforms such as Zoom to create text transcripts of meetings. Once these transcripts are in text form, AI tools can generate summaries of the meetings and even pick out a good share of agreed action items. Just as we did with the earlier generation of voice recognition assistants, we will increasingly take these tools for granted as a minor boost to our productivity.

### AI for Program Operations

However, building AI into your nonprofit program's delivery is a different matter. Building custom or semicustom AI solutions is expensive. Off-the-shelf products such as ChatGPT and its competitors are rarely directly useful in building a program or meeting a social need at scale. To create a significant new application of AI technology typically costs hundreds of thousands of dollars, and then similar sums are required for maintenance during the period that application is being used. As for any major investment, you need to seriously evaluate the cost-benefit equation. You may find that the projected benefits of an AI solution often don't justify the costs.

I have seen numerous nonprofits spend hundreds of thousands of dollars (or even more than a million) on AI solutions. Some of these solutions failed. Just because it's AI doesn't mean it will work. The same points I make about tech in general apply just as much to AI (or more). AI is just another technology that may or may not work in a specific real-world application.

Even higher costs are frequently associated with outsourcing AI technology work. Few nonprofits can afford to recruit and retain their own senior software developers or data scientists, so they hire outside for-profit consulting companies. But an hour spent by a developer working for a consulting company is typically at least twice as expensive as an hour spent by a developer who works inside a nonprofit organization. Of course, this assumes that a nonprofit knows how to hire and manage such a developer and can commit to employing them over time, something that is organizationally and financially beyond the typical nonprofit's means. This is one reason why tech-for-good nonprofits should exist for common needs in the social good sector: it is far more practical to spread the cost of tech talent over dozens or hundreds of nonprofits that need a similar solution.

It is possible for even a small team to build exciting applications with AI. For example, AI is very good at recognizing patterns in images, and a single experienced software developer can use the available tools to make this doable. The field has come a long way from a desktop scanner recognizing machine-printed text! If people can look at images and recognize something useful (a malaria parasite in an image from a microscope, a tumor in an X-ray or MRI image, a road in a satellite image, an endangered animal captured in a photo by a trail camera, or a fungus on a stalk of corn in a photo taken with a mobile phone camera), a social good application is likely possible.

For predictive AI, tech tools are available to software developers that can make it much easier to create a new AI application. For example, both Google and Facebook released their respective AI tools, TensorFlow and

PyTorch, as open-source tools available for anyone to use. These tools make it much easier for an individual software developer (or a small team) to develop AI applications based on data collected by their organizations. It is logical to expect that the coming years will see similar tools for generative AI solutions based primarily on data associated with a specific social good issue.

The for-profit world will quickly deploy technology where there's plenty of money to be made. The quality of AI products is going to get better and better, which will lower the cost of adapting AI to socially useful applications. My advice to the typical nonprofit leader about advanced AI technology is to wait until the price-performance of AI products gets truly appealing. However, waiting is not a great option for some bold nonprofits that see a great opportunity to apply AI and are willing to invest the money to do the hard work of applying it using the leading-edge tech. These trailblazers are how the field of tech for good advances.

Many of the tech social enterprises mentioned in this book are using AI in their products and programs and thus doing that hard work. For example, Benetech is busy using technology it calls "PageAI" to go well beyond recognizing text in textbooks to help students with disabilities like blindness. The company wants PageAI to be able to scan math, chemistry, and other equations (generally showing as pictures in books), figure them out, and then read them aloud (or provide them in braille). AI also holds significant promise for describing graphs, diagrams, and even photographs, tasks that are economically impractical for human labor to perform. Bookshare has millions of images in its accessible library, and it doesn't have anything like the funding to manually have someone describe the contents of each picture.

An example of a nonprofit that uses AI to solve a social problem that centers around images is Thorn, a Skoll Award–winning social enterprise that has created tools for combating the presence of child sexual abuse material (CSAM) online. CSAM are pictures and videos that depict children being abused sexually, and it is illegal worldwide. However, it is difficult to control. Traditional approaches to detecting CSAM focus on checking files against a list of digital fingerprints of known CSAM: images and videos that have been produced and seen before. However, new material uploaded daily eludes this approach, so Thorn built a CSAM classifier that uses AI to help online platforms (such as Flickr, the photo website) find new CSAM that has never been seen before. That way, there is the possibility of finding evidence of active child abuse and intervening to save a child from further abuse, which actually happened when Flickr discovered such images using Thorn's tool. In addition, this technology is also able to spot some of the new wave of deepfake content that uses AI to generate new and unique ways to depict terrible violence against AI-generated children (though typically these are trained on content depicting actual violence against real children).

Beyond recognizing text and patterns in images, AI systems have been getting quite good at translating languages. Many people have used their smartphones to translate signs or menus while traveling, another practical use of AI tools.

An example I find exciting is TalkingPoints, mentioned in chapter 4. Founder Heejae Lim's vision was to help improve educational outcomes for immigrant children in the United States by making it easier for parents and teachers to communicate about their children/students when the parents and teacher do not speak a common language. It's a great example of taking off-the-shelf translation AI technology and improving it to work better for the specific TalkingPoints education application. For example, when a teacher mentions a "sub," the program translates this based on the concept of substitute teacher rather than a submarine coming into class!

**MapBiomas: AI for Land Use**

MapBiomas is a Skoll Award–winning social enterprise developing image-based AI. It uses satellite data and AI to help people understand what's going on in land use in their countries over time and to help scientists, conservationists, law enforcement, and policymakers understand current and historical land use / land abuse patterns. It also has widespread use in research into climate change, in investigations into climate-related crimes such as illegal deforestation, and in motivating change.

MapBiomas was founded in 2015 by Brazilian scientists who were passionate about the climate, greenhouse gases, and deforestation. They came together based on a challenge: Was it possible to come up with comprehensive land use maps for all of Brazil that show the historical changes in how land has been used over decades? For example, was it possible to see what percentage of the Amazon's rainforest had been converted to cattle pastures or soybean fields?[14]

Their timing was good on the data and technology fronts. The US Landsat program had been collecting satellite data about the entire earth for thirty years, so the data needed to meet the challenge of surveying land use had become available without the need to pay for it. The resolution of the data was thirty meters, which meant it was feasible to distinguish from the satellite pictures and other data whether a given spot of land was rainforest, farm fields, pastures, a waterway, or a human habitation, such as a village or a city. Google Earth Engine was a powerful and free cloud tool available at the time (and still today) that made it feasible to create and deliver the planned maps to just about any user with a device and a data connection. And finally, it was possible to teach AI algorithms to recognize different land use types by having knowledgeable people label different areas by different types so that the AI could use the satellite data to replicate those labels for data covering an entire country.

As a result, the MapBiomas team was able to create time-lapse maps showing how land use in Brazil had changed over decades. Against a backdrop of the deforestation of the Amazon rainforest, these images were very powerful. One could watch how year by year land was converted from forest to rangeland for cattle and cropland for growing corn (maize) and soybeans. Over time, the team created new products based on this technology, such as MapBiomas Alerta, that could detect illegal deforestation events in nearly real time. Not surprisingly, the Brazilian scientists' research at the time demonstrated that more than 99 percent of the deforestation they were detecting this way was illegal.

The team took a very generous and open approach to their AI technology rather than a proprietary attitude. They made their software available as open source, so that other people could use and improve the technology. With MapBiomas AI technology, it is very easy to do your own research, analysis, and visualization work using Google Earth Engine, without even needing to upload any satellite data. This generosity has made it possible for new countries to replicate their own MapBiomas capabilities because all these tools make it much easier to get started.

The effectiveness of the project involved not just open-source software but also the fact that the maps were available to the general public, who could report errors in the maps or the AI outputs, thus allowing the maps and the technology to be improved. As the MapBiomas team writes in a major scientific paper, "With open access data, it was possible to perfect the LULC [land use and land cover] maps with end-user feedback, which reached one hundred thousand users in 2019."[15]

The impact of the MapBiomas project rapidly expanded. The scientific community had access to new information and new tools. Scholarly articles written by MapBiomas collaborators have been cited thousands of times in other papers because of the definitive information it has provided. The Brazilian authorities opened more than ten thousand investigations of illegal deforestation based on the information in maps generated by MapBiomas.

The innovations developed by MapBiomas have applicability well beyond Brazil's Amazonian rainforests, spreading to fourteen countries in South America and to Indonesia. In each new country, the projects are wholly driven by researchers and technologists based there. The AI algorithms are widely shared but need to be adapted to the unique circumstances of each country. The open-source nature of MapBiomas makes this possible.

The issues now being addressed with the MapBiomas approach are also wider than just deforestation. To truly understand land use, it's important to understand issues such as mining and urbanization. Technical teams are also working on projects focused on fire, water, and soil (including carbon content).

Climate change is a vast and complex human and planetary issue. Part of any systems change initiative is knowing the current state of affairs and having a way to track progress. The MapBiomas team took the challenge they were first given in 2015 and found a way to bring much-improved information about current and historical land use to both the general public and interested leaders. MapBiomas is a shining example of using a portion of the vast amount of geospatial data being generated by satellites circling the earth to create a powerful digital public good.

## QUESTION 4: WHAT SHOULD WE DO WHEN THE AI GETS IT WRONG?

Beyond the cost to develop AI solutions and the ethical questions about collecting and using sensitive data, there are the ongoing costs of maintaining a deployed AI system. Every tech-for-good system has these ongoing costs, especially if the system is connected in any way to the cloud.

However, one of the most underappreciated aspects of deploying AI solutions is the cost of remedying the mistakes the AI solutions make. They all make mistakes. Some of these mistakes are due to the bias problem we've already discussed, where the datasets used to train the AI are not well matched with the application and the community being served. Sometimes errors are simply intrinsic to the limitations of AI algorithms: they are never perfect, and there are applications where they don't work at all well. After all, most social problems are not well matched to the capabilities of today's AI software!

There is now a tendency in society to believe computing machines are intelligent and somehow smarter than people. As I pointed out earlier, AI systems aren't actually intelligent. They don't know when they are wrong, even when they "sound" supremely confident in presenting the information they produce! AI errors show up in every way imaginable. Even optical character recognition, which has been around for decades, makes mistakes in recognizing individual letters. Voice-driven systems misunderstand commands. AI systems analyzing medical images can label something benign as a problem (a "false positive") and miss a deadly problem completely (a "false negative").

The first way to deal with AI errors is to take a cost-based approach when assessing a new application of AI. Imagine a system where a team of people are doing a task again and again. AI is pretty good with repetitive tasks. If an AI tool helps that team to do twice as many tasks in the same amount of time, including the time they spend fixing mistakes created by the AI, that's probably a good application of AI as long as the mistakes don't increase (compared to those humans would make) and the cost of the technology is small compared to the human labor costs.

Sometimes the cost to fix the mistakes outweighs the benefits of using AI tools. For example, my original for-profit optical character recognition company, Calera, figured out that recognizing 95 percent of the characters on a page accurately wasn't useful in large-scale data entry with a 99.9 percent accuracy target. It turned out it was quicker for a human to type in all of the text on the page than to painstakingly have a human operator correct each of the errors the algorithm had made. Even though Calera theoretically was doing 95 percent of the work, it didn't turn out that way once the cost of fixing the errors was included. To be useful for data entry, Calera had to deliver significantly higher accuracy before it saved enough money to justify deploying its AI-powered tool.

The more powerful the AI system, the more trouble its errors can cause. The cost of AI mistakes can also go beyond monetary damages. Reputational damage can be devastating. A great example of this kind of trouble happened to the lawyer mentioned earlier, who took a shortcut with ChatGPT, got caught, and was legally sanctioned for acting in bad faith. The people using AI tools can't abdicate responsibility for their jobs: saying "The AI made me do it" is no defense.

As a social innovation leader, you need to be looking at a new AI application not only in terms of its cost-savings financial case but also in terms of the risks of a catastrophic reputational hit, which could sink you and your organization.

This is why I generally recommend against using AI tools for high-stakes, life-or-death issues. Mental health, physical health, and safety are areas where you should proceed with caution. Above all, you need to maintain humans in the loop to oversee the system to prevent bad outcomes.

A great example of the bad outcome kind was the use of AI by the National Eating Disorders Association (NEDA) in the United States. During the COVID-19 pandemic, calls to NEDA's hotline exploded. Its counseling staff felt overworked and underpaid, and by 2023 the staff started an effort to unionize. Not long after that, the human counselors were fired, and a new generative AI chatbot, Tessa, was put on the helpline.[16]

Tessa lasted a week. An advocate for people with eating disorders circulated screenshots that showed Tessa advising them to count calories and go on a diet, which is exactly the opposite of what modern weight disorder advice would be because the "just lose weight" messaging typically makes things worse for people with eating disorders. This is one of the risks when deploying an open-ended generative AI solution without human oversight: if Tessa's output was trained on what's found on the modern internet, we should not be surprised that it produced bad advice!

NEDA's response was to blame other people. Its management said that the outsourced chatbot vendor was at fault and that the advocate who circulated the screenshots had worked hard to get Tessa to say what it said.[17] As one might imagine, this defensive approach was suboptimal in a public relations disaster. In the end, the organization apologized for failing the community it existed to serve. However, NEDA's experience is an object lesson in the risks of rapidly applying AI in a high-stakes situation.

The foregoing examples examine cost from the standpoint of the organization's cost structure and reputation. But ultimately the most important consideration is the impact of mistakes on the individuals and communities served by social impact organizations. NEDA's reputational disaster stemmed from giving bad advice to people who had reached out for help but could be hurt by that advice.

What if an agricultural advice solution looks at photos of a poor farmer's diseased crop and misdiagnoses the pest? What if the farmer spends their savings and borrows money to buy the suggested remedy, and it doesn't work?

The loss of a crop for a subsistence farmer may have a disastrous impact on that farmer and their family. High-quality nonprofits in the agriculture field take very seriously the possibility of giving the wrong advice.

Even beyond the communities meant to be served, AI tools are now introducing horrible problems for people who aren't even using AI! Stanford published a study of software programs that claimed to detect when students were turning in essays generated by AI.[18] These tools mistakenly labeled "only" 10 percent of the essays written by native English speakers as AI generated when they were written by a human but labeled more than 50 percent as AI generated when the essays were written by nonnative English speakers. Yes, the AI detectors, powered by AI, incorrectly flagged more than half of the essays actually written by nonnative English speakers (humans) as machine generated. You might be tempted to say, "Bad robot," but what about the professor who chooses to fail students based on the output of a worthless AI tool? That professor is outsourcing the high-stakes decision to label someone a cheater to a brain-dead piece of software. I'd say, "Bad professor!" We all need to make sure we aren't accidentally falling into a bad professor trap with AI.

Humans are, of course, fallible, and human society has evolved ways of dealing with the kinds of mistakes that people make. One of the goals of a well-run bureaucracy is to minimize errors or, at least, to minimize the cost of errors. Entire fields are dedicated to managing mistakes, whether in the testing of new drugs, quality control on an assembly line, food safety testing, engineering standards for buildings and bridges, or entire types of insurance products. However, these systems have been designed to catch the kinds of errors *humans* make.

AI algorithms make new kinds of errors in addition to the same kinds of errors people make (since the AI is often trained on human-generated datasets). Society, including organizational leaders, are not used to these new kinds of errors. They mistake generative AI results for search engine results and confidently believe the AI when it cites facts that don't exist.

Humans are used to existing patterns of errors but very sensitive to a new kind of error—errors that a human would not make but an AI system would. This is similar to the asymmetric dynamics of a vaccine. A vaccine could save a million lives, but if it kills or disables one thousand people who wouldn't have been killed or disabled by the disease involved, many people will be reluctant to take that bargain.

The European Union, as usual, has gotten out in front of these issues with its AI Act of 2024.[19] The act breaks AI applications into groups categorized by risk to human beings. It starts by outlawing AI systems that pose an "unacceptable risk," such as manipulating vulnerable people (i.e., children) and systems designed to classify people, including social scoring, biometric analysis, and real-time facial recognition. The outlawing of classification systems is not surprising given the publicity China has gotten for its social credit system and image recognition products designed to identify and track members of the disfavored Muslim Uighur community. "High-risk" applications of AI, such as self-driving cars (and planes), are subject to existing product safety rules. Other risky applications in fields such as employment, education, and law enforcement will need to be registered and assessed before and after being launched. "Lower-risk" applications will need to disclose to users that they are "interacting with AI." Finally, some applications are labeled as having "minimal risk," such as AI-driven spam filters, and thus are minimally regulated.

Of course, the possibility of something going wrong exists with every new endeavor and every new year. My goal in sharing what can go wrong with AI applications is to ensure that you are weighing the downsides as well as the upsides. Taking a risk-based assessment approach (as the European Union is) is a great starting point. Most of the problems with AI-caused errors can be addressed with planning and thoughtful design. If the cost of mistakes can be managed, and the new tech tool still makes outstanding mission and financial sense, you might just have a winner.

### QUESTION 5: WHAT PROCESS SHOULD I FOLLOW WHEN CREATING A NEW AI APPLICATION?

Assume you are ready to build a major AI-powered tool. You didn't start with AI; you started with the problems you want to solve. You talked to users and found an important and pervasive issue. You believe you understand how to make a positive impact on their lives with this new idea, and a solution based on AI seems more likely to scale impact than the status quo or other tech options. In addition to the advice in the rest of this book, you will need to consider particular issues that concern AI tools.

First, you should be able to tell a compelling story that links the problem you are solving to the tech solution you are proposing. Next, you should see if you have enough data. It's quite common to need data to train the AI tool and test it before you deploy it. Once you launch the tool, you will need to know if it is working. We'll talk about the challenges around datasets later in this chapter, but AI runs on data. If you have minimal data, your AI applications are going to be limited to off-the-shelf solutions, which are based on datasets that may not be applicable to your use case(s) (and might make things worse).

Think carefully about the humans who will be using your proposed solution. Will it make their lives easier and better? Will it solve a problem they have? If your new tool is going to be used by your staff, have you done the legwork to ensure that the team will enthusiastically welcome it? Many new tech ideas fail on contact with nonprofit staff members and volunteers who are not convinced about the need to change the status quo.

You might want to try building a human-powered prototype that simulates what you have in mind. If nobody likes your solution when it's done by an expert human (pretending to be a machine), they are probably not going to like the solution when it's done by an algorithm. A major AI project involves spending a lot of money, so adopting an agile approach to development allows room to revise the plan based on what you learn. You may even need to put the project on hold if it seems like the original story or even a new and improved story is unlikely to come true.

What's your plan to deal with the mistakes the AI makes, especially costly mistakes? One of the easiest ways to manage mistakes is to deploy AI tools that make humans more effective in doing something they are already good at. If the AI makes a mistake, the human in the loop should recognize and fix it. When a search engine or a voice assistant doesn't give us a good answer, we adapt and ask the question differently. Will something like that work with your tool? Perhaps the mistakes will lead the humans in the loop to come up with a recommended improvement to the tool.

Many AI-based tools build in an entire AI mistake-improvement cycle. For example, one of the original successful chatbots for social good was Roo, the sexual health advice app created by the tech team at Planned Parenthood. It was quite successful because it worked only on questions related to sexual health and because many young people were more willing to ask a robot what they considered embarrassing or sensitive questions. However, a regular improvement process was also going on monthly, where expert health educators were reviewing questions that Roo stumbled with and improving Roo's answers. It wasn't realistic to simply deploy a complex chatbot and then expect it to work without regular and constant improvements.

### QUESTION 6: HOW SHOULD I APPROACH GETTING THE DATA WE NEED?

An AI solution is only as good as the data used to train it. Just about all of the commercial AI tools are trained on datasets limited to the data that the AI-development companies could find (or, as many publishers have claimed in lawsuits, steal). These datasets are not representative of all of humanity, however. Far more data from wealthier and technologically advanced countries and in English are used to train AI, even though the populations of these areas represent only a fraction of all humanity. Minority populations are systematically underrepresented. Even though women are the majority of the total population, they are systematically underrepresented in data.

If you ask a generative AI application today for an image of a judge or a CEO, you will probably get a picture of a man. According to *Bloomberg*'s article "Humans Are Biased, Generative AI Is Even Worse" (2023), when the authors asked for images of "a judge," only 3 percent of the results showed a woman, even though a third of judges in the United States are women.[20] When the researchers asked for an image of a fast-food worker or a social worker, the AI would produce a person of color even though two-thirds of the people in these professions are white. The AI algorithm didn't have these systematic biases encoded in it to start with: it acquired these biases by being trained on content that demonstrated the biases. Such biased algorithms are like predictive policing, which is likelier to dispatch police to neighborhoods of people of color than to white neighborhoods.

These built-in biases are problematic for social change organizations intending to remedy society's problems. The biases are insidious: you don't know when a mainstream AI model might inappropriately encourage a young Black man to become a janitor when he's better suited to be a professional or screen him out from a list of interviewees. While Joan Mellea and I were at Benetech, we did research on the AI tools already being used in recruiting and hiring in 2018 and concluded that almost all of them discriminated against people with disabilities.[21] If you were to replace "people with disabilities" in this sentence with "women," "people of color," "immigrants," or just about any other minority group, the problems are likely to be the same.

It gets worse. Even though the generative AI companies claim that the data used to train their AI models are not retained in the models, these claims have been repeatedly disproved by researchers (and by the *New York Times* in its lawsuit against OpenAI and Microsoft).[22] Images remarkably similar to copyrighted works, including the artist's signature, have been produced by generative AI image generators. A leading open-source image generator tool, Stable Diffusion, was infamously trained in part on pictures of children being sexually abused.[23]

It is very clear that social change organizations should not be feeding confidential information from their clients into these commercial platforms. It wouldn't be ethical to do so. You might want to have these tools help you write a nonconfidential grant proposal as long as you are OK with that platform advising other nonprofits based on your text (and the text of a host of other nonprofits). But if you want to use AI tools on confidential client and user data, you need to keep your data out of datasets used by the for-profit tech companies. Of course, the need to protect this confidential data is important for many other reasons.

Crisis Text Line is a cautionary example of what happens when a nonprofit shares confidential data with a for-profit. Crisis Text Line has been a pioneer in the field of using text messages instead of phone calls to respond to young people in crisis. It received hundreds of millions of text messages and was known for using that giant database for improving its programs, including counselor training to enhance empathy and deliver better services. These insights were so valuable that Crisis Text Line started an affiliated for-profit company to apply these insights to commercial call centers.

Unfortunately, Crisis Text Line shared text transcripts from its helpline with this for-profit company, although the transcripts had been scrubbed of personally identifiable information.[24] When the news broke that a suicide prevention helpline was sharing the confidential chats with its clients with a company trying to make money, the reputational damage to Crisis Text Line was significant, even if it quickly terminated this data-sharing agreement.

It is easy to look back and see this sharing of data as a bad idea, but that wasn't so clear when this project was initiated. Generating revenue to support the nonprofit mission is something that most tech-for-good nonprofits think seriously about. I feel confident that the board and staff at Crisis Text Line saw this affiliation with a for-profit as an opportunity to raise needed funding for their important counseling work. However, this instructive episode has served to change the minds of leaders in mental health and related fields about what uses of data are allowable in the future.

The importance of data and building datasets will only grow in tech for good. If we are going to use AI to tackle many of the world's big problems, we are going to need an ethical revolution in the data-related capacity of the social change sector. We need to collect the right data, secure them, scrub them of personally identifiable information, and use them to benefit the communities we serve, not the owners of large tech companies. In chapter 8, I discuss these opportunities and propose a set of best practices for collecting and handling the private data of the people we serve.

## USING AI IN SOCIAL GOOD FOR HELPLINES

Helplines are a great example of a field that is increasingly benefiting from AI. The big tech advances in the field in recent years have stemmed mainly from helplines branching out to assist young people via text in addition to by telephone. The Aselo platform at Tech Matters is trying to make it far easier for helplines to add text channels such as WhatsApp and webchat. Text-based conversations are more amenable to AI applications, and I want to share two tech efforts in the helpline field that I admire.

The Danish child helpline Børns Vilkår (Children's Welfare) made a philosophical decision not to have its AI tools interact directly with young people. It instead developed an AI "Counselor Guide" to assist volunteer counselors.[25] The helpline's existing knowledge database was available to its counselors, but in practice it wasn't easy for volunteers to search the database while on a text conversation with a child. The solution, the Counselor Guide system, pays attention in real time to the text messages being sent by the child and tries to identify the key topics being discussed (e.g., depression, bullying, impending divorce of the child's parents, and so on). Based on that assessment, it presents some short and useful advice about these topics alongside the text conversation to assist the volunteer in doing a better job of counseling. The Counselor Guide gets some of the topics right in the majority of conversations, and the volunteer counselors like the support it provides. And if it identifies the wrong topics, the counselors are smart enough to know that, and they can always search the full database if they want more information. Straightforward, helpful, but not magical: this is what a practical, values-based application of AI looks like.

Another helpline using AI in a sensible way is the Trevor Project. With support from Google.org, the organization developed the "Crisis Contact Simulator" for training their volunteer counselors.[26] Volunteers go through a simulated role-play version of a conversation with a teen in crisis. After completing two of these sessions, new volunteers move on to training sessions with expert human trainers. The benefit to the Trevor Project is that it could train more volunteers with a given number of expert human trainers. It also helps the expert human trainers because they don't have to role-play being a suicidal teen every day, which is a very stressful task. If the AI or the volunteer makes any mistakes in the training sessions, it doesn't affect a real teenager in crisis. It is a learning opportunity. Some of the team who developed this role-playing capability have spun out of the Trevor Project to create the new social enterprise ReflexAI to provide this technology to other helplines.

These two examples show how AI in the form of algorithmic tools is helping two specific crisis response helplines. Instead of replacing humans, the tools make the humans in the system more effective. These innovations are already being adopted by other helplines.

## RECOMMENDATIONS

AI is an incredibly powerful technology. It represents many of the advanced uses of the data gathered by tech tools. However, powerful technology comes with challenges. It's expensive to work with AI, and the cost of the mistakes it makes may sink your enterprise both financially and reputationally.

If you want to build a tech tool that collects data, you should plan ahead. If you are going to hold onto data for possible AI applications, be sure you've obtained those data ethically. Be transparent about how you will use the data and promise not to sell out the interests of the people who are the subjects of the data. Protect the data. Scrub them of personally identifiable information if that applies.

When you're ready to build an AI-based tool, treat it like any other tech tool. Design it with the social problem and the users' needs in mind. Keep humans in the loop. Be agile. Experiment. Try things and learn from what does and doesn't work. Don't rush to deploy something that might have costly or fatal error patterns.

Many of the most exciting tech-for-good applications in the next decade are going to use AI. I hope your ideas are among them!

## CONCLUSION

AI technology has made steady advances over the years, and leading nonprofits are making active use of AI. However, the most successful organizations already have extensive experience in collecting and using data to make their efforts better, and AI is a logical next step for these data-informed teams. Off-the-shelf AI products right now are best deployed on internal administrative tasks such as writing, editing, and fundraising. At this point, more capable AI applications that improve program activities almost always require a technology development team. Such teams are not inexpensive, so the benefits from such an investment need to be substantial. For many programs, waiting for an AI product to be developed is probably the best strategy, especially for the nonprofit without deep data or technology experience. The best source of advice as you evaluate AI projects are your peers and the stories about what did and did not work for them.

# 6 THE TECH-FOR-GOOD LIFE CYCLE

Outmoded technology and outmoded design approaches hobble the social sector when it comes to the effective application of tech. However, the social sector's focus on short-term tech pilots and one-off projects is what truly sets tech in social impact behind. One of the most essential pieces of advice I give nonprofit leaders about technology is not to think of tech as a separate project. They should instead think about creating a sustainable social enterprise that uses tech to achieve their mission over many years. Whether the technology they are building is for a purely internal audience made up of their team or an external audience of community members or other organizations, people will be depending on this solution to be there for them.

Why is this so important? First, in the for-profit world no budding tech entrepreneur quits their job to start a tech project. Instead, they start a company, an enterprise, that is going to deliver outstanding customer and shareholder value over time. Second, in the for-profit world no company in its right mind will choose to install a new accounting system or a CRM system with the idea that the company will stop using it in two or three years. And yet the nonprofit sector seems to be full of such short-term tech projects.

When the team at Tech Matters surveyed people all over the world as part of the startup process for our Terraso platform, new technology for local leaders struggling with climate adaptation, we asked our potential users what tech tools they were using. We heard about hundreds. When we looked at this long list, we realized that only perhaps a couple dozen of them were likely to still be around and actively maintained three or four years in the future.

Not surprisingly, the tools with a longer-term sustainability model were far more likely to have happy users.

That's why the central message of this book for leaders of tech-developing (internal-use) and tech-providing (external-use) efforts is to think about creating a long-lived tech effort. Systems change does not happen in one year or even three years. If your technology solution is going to have a major impact on your organization or your field, you should be thinking in timescales of a decade (or, more likely, two decades) to reach scale and deliver serious change. Longevity is essential when building core technology for a field or an ambitious organization. People need to believe that the tech will exist for a long time before they will base their work on it. It's not enough to build a project that delivers tech for good for only a year or two; tech-for-good nonprofits must be built to last. Of course, this advice is also directed to nonprofits that do not think of themselves as a tech group as long as delivering their mission depends on core technology for their team.

Luckily, the financial goals of a nonprofit tech enterprise are much easier to reach than those of the for-profit startup entrepreneur. Nonprofits only need to find the money to pay their team because they don't need to deliver massive profits to satisfy venture investors. They don't even need to pay back their funders! Technology teams are, relatively speaking, more expensive than program teams without tech specialists, but it's completely conceivable to have a successful tech-oriented nonprofit with an annual budget of $500,000 or $2,000,000. A $2,000,000 nonprofit that breaks even and revolutionizes a field is a giant success by nonprofit standards. That's moving the goalposts a long way!

Motivation differences between for-profits and nonprofits aside, there are many similarities between the two that have an impact on sustainability. Both have the same phases of development: exploration, initial development, growth, maturity, and exit. However, before we explore these phases, it is time to talk about which legal form your enterprise should adopt.

## NONPROFIT OR FOR-PROFIT?

If you are starting a new technology organization for social good purposes, the question of whether to incorporate as a nonprofit charity or as a for-profit company will arise. Readers who have already set up their organization as one or the other may want to skip ahead to the next section in this chapter—unless, of course, you are reconsidering your choice of organizational form.

I do not believe the choice of organizational form is a moral issue, although it often feels that way. When I talk to founders of a new social enterprise, the second-most popular question seems to be about the nonprofit/for-profit choice. The most popular question is, of course, "Where can we find the money for our new effort?"

The question of organization form was sufficiently frequent that back in 2011 I wrote an article entitled "For Love or Lucre" for the *Stanford Social Innovation Review*.[1] My main message was that organizational form was one of the *last* questions you should try to answer. I suggested that new social entrepreneurs first think hard about their personal motivation. If getting rich is your biggest goal, providing products and services to people who can't afford them is probably not your top option! If you care about a social cause or a community of people above all else, then you can give starting a nonprofit serious consideration. After you have worked through questions of motivation, market fit, capital requirements, and your attitudes toward control, the choice of nonprofit, for-profit, or something hybrid should be relatively clear.

However, there is a reason that 95 percent of the case studies in this book are of nonprofit organizations. I've been a Silicon Valley for-profit tech entrepreneur and had my social good idea vetoed by my venture capitalists for "excellent business reasons." I am deeply skeptical of the claim that you can be a for-profit and put social impact first. The choice of corporate form is a choice of what comes first in your decision-making. If you are a for-profit, and you have taken capital from investors, you are obligated to make them money. It was completely fair of my venture capitalists to demand our for-profit company focus on making profits, making them their investment returns, because that is the promise we made when we took millions in capital from them.

I have seen too many unhappy social entrepreneurs who were talked into becoming for-profits. There inevitably comes a day in every organization's life cycle when difficult choices need to be made. A for-profit typically will pivot toward more profitable markets when faced with a strategic choice, jettisoning or much reducing the priority of the social good agenda. The founder who thought that they could make returns and do extensive social good find themselves focusing on financial returns instead of on social good. This outcome can be quite disheartening for a leader who wants to do more than just make money. "Doing well by doing good" will most likely end up becoming "Doing well at the expense of doing as much good as you had hoped."

As the founder of a nonprofit, you explicitly put social mission first and money second. By itself, doing that doesn't make funding magically appear! However, you have the clarity that your goal is to make the maximum social impact with the money you can get your hands on. For a nonprofit that provides its products to external customers, you might plan that your social enterprise reaches 100 percent funding of your expenses from customers who can afford to pay for your products. But what happens if you reach only 80 percent revenue support, while having outstanding positive impact? A nonprofit can always go to donors and make the pitch that the donor money is matched five to one! In my own experience, organizing as a nonprofit meant that even when we were struggling financially, I always knew that if our work was otherwise making a real difference, I could go to the donor community and probably find enough money to make it through the tough times.

Over my long career, I have seen many for-profit companies making positive social impact. The assistive technology industry was initially made up of many entrepreneurs who cared deeply about building tools for people with disabilities. Many of these entrepreneurs had a disability or had a close family member with a disability. They all could have gone into other industries with their skills and probably made much more money but chose to stay committed to serving people with disabilities. Most of these entrepreneurs were self-funded and didn't have to answer to investors. Their goal was to make a decent living for themselves and their families while doing something important. But being a for-profit business still meant that they had to focus on serving people with disabilities in the United States or perhaps Europe, rich economies where expensive technology could be supported by the state, employers, and families. These for-profit entrepreneurs were choosing to help a disadvantaged group, such as people with disabilities, but couldn't afford to address the needs of the 90–95 percent of people with disabilities who live in less wealthy countries. Their business model didn't work for the vast majority of the people who needed their products. It's one reason why I started Benetech as an assistive technology company organized as a nonprofit.

Whether your tech-for-good vision involves starting a new organization or adding a major new strategic tech program to an existing one, here are the phases you will likely go through.

## THE EXPLORATION PHASE

You have a great idea for how to use technology for social good. You probably have a social problem in mind, if you don't already have an organization with a specific social mission. Who will benefit from your new innovation, and why do they need it? It could be that you know all about the social issue and the need, or it could be that you just imagine the need and want to help. Perhaps your organization has worked in a field with a certain need for a long time and wants to do something new because the old ways aren't meeting the need.

Before you get to work on your new idea, there's always the question of where the time and the money for the new startup might come from. If your project is in an existing organization, you might be able to use some of its people and funding to do your initial investigation. If you are starting something from scratch, can you find a donor who is comfortable with risk? A fellowship to explore a new idea? Or can you do it on your own time, building up a case to help you get funding later in the process? We'll talk much more about fundraising in the next chapter.

The next question should be: Am I (or are my team and I) on the right track with this idea? If we build this thing, will people use it? Will it make a big difference in their lives? As noted in the previous chapter, the first step on this path is to talk to your potential users. I mean *your* users. In the nonprofit sector, where many of the decisions are heavily influenced by donors and senior leaders, so much of planning something new can be directed at people who will not actually be using the tech you're developing. Given that you're working in the social sector, you will need to engage senior leaders and donors, no matter what, but be clear that although their support is helpful to your project, it is not sufficient to assure the project's success because that depends on the actual users.

When my team starts exploring a new project, we emphasize the need to speak with twenty to thirty diverse potential users as the initial step. If you are building a new tool for an internal team, you have a captive audience to engage! If you are new to a field, you may need to depend on existing nonprofits to help you find potential users because most potential users typically have no reason to talk to technical people: they don't know about such things or what they know makes them reluctant to talk. You must count on being introduced to the potential users by someone they trust as a key part of increasing the chances of getting useful and actionable answers. The need to generate and maintain trust is a crucial part of the entire life cycle of a successful tech-for-good effort.

My team and I like to ask open-ended questions when we talk to potential users, such as

- What are you trying to accomplish?
  - Here we are trying to understand what purpose(s) the potential tech solution would be supporting because tech should not be an end in itself.
- What's working well with your current technology?
  - If we build something new, let's be sure not to accidentally make things worse by leaving out something that is currently working just fine.
- What would be "even better if" with new technology?
  - This positive framing, rather than asking what's wrong with the current tech, can minimize defensiveness about broken tech tools or past mistakes.
- If technology could do anything for you and your family/community/organization, what would it be?
  - You probably can't deliver technology bliss, but knowing what your potential users think tech nirvana looks like is always helpful.

At the nonprofits I have helped found, we make an effort to interview people who represent the full range of potential users. Since we are usually trying to build a product that will meet the needs of users in both emerging and developed economies, we need to interview potential users on every continent and at the full range of resource availability. We want to talk to a few users who have money but to many more who don't.

It's not always possible to interview the potential external users directly. The executive director of an NGO may not be comfortable with you speaking directly with their staff or the people they serve. Many nonprofits work with people in crisis, and it's often impractical (or even unethical) to use the time of suffering people for your initial investigations. You might look initially for interviews with people who are proximate to the issue and target community—as close as possible—and whom you can access. For example, a field staffer who works all day with the target community is a better proxy for the tech design interviews than someone stationed in headquarters in Washington or Geneva!

By the time you've done twenty or so interviews, it's pretty clear if you have an exciting opportunity for a tech enterprise. Not surprisingly, the needs you discovered and your ideas for potential solutions are likely to be different than what you originally thought or what your trusted partners or donors had in mind. If you focus on the actual expressed needs of the user community, the odds of success will significantly increase. In a world where most tech projects will fail, you want to tip the balance in favor of success. In an agile design process, you will also have a good idea of where to start in demonstrating value.

Of course, I should point out that most of my early project ideas did not succeed. Your mileage may vary, but lowered expectations for initial ideas are in order, especially for ideas that start with a technology to use as opposed to a social problem to solve. As the longtime tech-for-good leader Ann Mei Chang points out, you should fall in love with the problem, not with your solution.[2] The most common result of tech-first approaches is that the proposed idea is not compelling, and nothing else was discovered that merits a pivot to something better. There are so many reasons why new tech ideas are not good ones, including:

- The problem you wanted to solve is not an important one to the real people in the communities you thought needed help.
- The existing way the need is being addressed is good enough. A new idea has to be much better than the status quo to justify changing the way people are used to doing things.
- The technology doesn't match the needs, doesn't exist yet, or isn't ready.
- The proposed product doesn't work well enough, or it requires infrastructure that is not available in your target geographies, or it requires too much training because of its current complexity.
- The solution is great, but it's not affordable.
- The solution is great, but you have no sensible way to get it into the hands of the people who need it.
- The amount of funding you need to create and sustain your solution isn't available.

These reasons are just a sampling: there are many, many reasons why the idea isn't right or the timing is poor or the people involved are not ready for the idea. Because we are focusing on technology for social good, ideas that require immense investments in technology development are unlikely to get funded. That means you probably should explore solving problems with technology that already exists and on situations where there is implementation work to be done but where the fundamental technical feasibility is not in question.

It takes courage to kill an idea, especially when you, your team, or even your donors have fallen in love with it. In the tech field, smart people, including powerful forces, believe exclusively that tech must be the answer. This widely held belief has been labeled as *tech solutionism* and *technochauvinism*.[3] For a responsible steward of charitable and/or social impact funding, and for someone who is ultimately accountable to the communities you hope to serve, it's far better to cancel a project (or put it on the shelf) when all you've spent is dozens or hundreds of hours, not hundreds of thousands in funding.

My first project in the climate field was killed at an early point. I had come up with an idea called CityOptions. A city leader would go to a website, enter in their city name and location, and then answer ten or twenty questions to fill in the gaps in the easily discoverable information already on the internet (such as the population of the city or its weather). My vision for CityOptions would be that out of the thousand things you could do in your city, it would recommend the top ten in terms of climate impact and fundability.

I talked to mayors and city managers, community leaders, donors, and policymakers about my exciting new social enterprise concept. The typical reaction was, "Great idea, Jim, that sounds useful. But can I tell you about the three things my city really needs?" After twenty conversations where my idea was always "great" but never a solution to a top-three issue for any city, I realized that CityOptions' time had not come. We canceled it and let people know. After CityOptions failed, I tried very hard not to fall in love with tech solutions I had dreamed up before I found out if the people who would be using them would also fall in love with them!

The experience of repeated failure encouraged me to come up with a stronger rubric for picking attractive ideas. The bare minimum is that the social benefits in terms of impact on the communities you serve must significantly exceed the costs. Many studies have shown that it is better to give people in disadvantaged communities cash directly.[4] Spending the same amount of money in creating a charitable program intended to help the same people might not be as effective. Doing something new may not be better than the status quo if the funding would otherwise go to effective current programs or directly to the poor. Admitting this isn't always easy, especially when donors in the field are excited about funding the latest shiny tech fad.

If you are part of an existing mission-based organization, and your exploration has located a great idea for tech that will help your organization deliver much more in results without a big increase in costs, you are definitely on the right track.

Our rubric at Tech Matters for starting a new social enterprise is more general because we are willing to work in many different fields. Our assumption is that we know how to build tech enterprises, but as generalists we don't know enough about a specific field without domain experts helping us. So my definition of a winning idea is one where:

- The leaders in the field itself are asking for our help.
- A reform movement is taking shape (because new tech in a locked field is not likely to succeed by itself).
- At least ten million people will benefit directly or indirectly from a new tech solution that meets an unmet need.
- The for-profit market will fail to build a product for these people, usually because the market seems too small.
- It seems likely that we can build a sustainable social enterprise in the $2–10 million per year budget range.
- We expect in five years to be able to get the majority of our budget from customer revenue rather than from donations and grants.

Not every tech-for-good project needs to meet these conditions, but they are definitely issues you should consider before committing to start a long-lived tech-for-good enterprise. Note that the tech itself is only one part of a much bigger set of criteria: it's rarely the gating factor for success.

You might have to explore multiple ideas in a given field before you find one that seems as if it could be a fit. Don't despair! There are many real needs where a tech solution could be just the thing. It's worth taking the extra time to find the tech solution that really can have a social impact.

## THE INITIAL DEVELOPMENT PHASE

Hurray! Your exploration phase was successful. You interviewed a representative sample of the people you want to help, and it seems that you have a winning tech concept. It's an idea that seems likely to deliver outstanding impact for the amount of time, money, and effort you are planning to commit to it. It's time to start the work of assembling the team that is going to create the technology to meet the burning need.

For a substantial tech development effort, you will generally need a product manager. At the most basic level, the product manager represents the voice of the user in the product development project and makes the key decisions about what is in the product. You don't want your development engineers to have the final say on the product because the engineers are almost never your prospective users and typically struggle to put themselves in the shoes of the user. Of course, having a diverse development team tends to reduce the gap between the technologists and the users, as mentioned in chapter 4, "Designing Tech for Good."

In an ideal case, you will have ten or fifteen users who are sufficiently interested in your product that they will commit to joining your product development journey. A practical reason for this number is that some users volunteering to help might end up being too busy to respond. Numbers like this are not practical for most internal projects, but if you have a lower number, their degree of engagement may be higher.

The goal is that these users helping the product development process represent the range of users for your planned product. I sometimes joke that the role of these codevelopment users is to regularly test "progressively less broken" versions of your new product. In the case of software development, we might release a new version of the software every few weeks. A hardware product will probably have a slower pace of new version release, but the idea is the same. Release early, release often.

If users see that you are making regular progress and listening to their feedback, the process is quite reinforcing to them. This is especially true when their past experience is of being ignored by technologists! Don't emulate the tech people who claim they know better than the users what is needed or assert that the tech they're developing is too complicated for the users to understand (and try not to have a tech person on your team who has these attitudes).

With a large project, the codevelopment phase might take a year or more. You will need to find donors who are willing to spend hundreds of thousands of dollars (or euros) simply on the belief that you will create something useful.

You will need to remain open to the idea that your initial enthusiasm was overly optimistic. At Benetech, we have had to kill a couple of projects over the years after investing hundreds of thousands of dollars in them. In both cases, I believed the world needed the product we were trying to build, but we needed to stop working on it. After starting these projects, we learned enough to realize that the odds of success had declined to the point where it no longer made sense to spend more donor money.

### Clearing Storm: My Best-Loved Failure

My best-publicized example of failure was Benetech's humanitarian landmine detector project. Landmines and unexploded ordnance (bombs and shells that might still blow up) continue to be a global scourge, killing and maiming civilians long after a war has ended. The social need is so important that I fell in love with both the problem and our idea. Our mistake was to think that we just had to build an affordable landmine detector that located explosives instead of detecting metal, which was the then current technology. Metal detectors find all kinds of metal, not just the small amount in a typical antipersonnel mine. On average, a metal detector has a ratio of one hundred false alarms to each true landmine detection alarm in an actual minefield! Wouldn't it be better, we thought, if we detected an actual landmine each time? We had a partnership with a technology company that had invented a novel explosives detection technology for the US military, and its tech team was very excited about their invention also helping reduce harm to innocent civilians. I was able to attract two top-notch technical leaders because of how inspired they were to work on this terrible problem. They named the project after a famous Ansel Adams photograph: Clearing Storm.

Unfortunately, Clearing Storm ended up needing to be shut down because of a problem we never anticipated: political risk. Landmine detectors are classified as an offensive weapon under US export law because they can be used to breach a defensive minefield in time of war. The laws are very strict: we needed the permission of the US Departments of Defense, State, and Commerce. Plus, we needed to license the technology itself from the for-profit company that invented it. We started working on the project, but within a little more than a year it became clear that we were not going to be able to get all these permissions at the same time. The parties were supportive of humanitarian demining, but during that period American soldiers were being regularly killed in foreign wars by improvised explosive devices. Even though our version of the product would have been too slow to be used in an offensive manner in combat, it was clear that the political risk of allowing the export of this technology was too high.

Our primary funder for the project was the Lemelson Foundation, which had a mission to encourage invention and innovation. Lemelson was more comfortable than most foundations with the idea of failure. We reluctantly laid off our two brilliant tech leaders and offered to return to the foundation the remaining unspent funding out of a $250,000 grant. However, I asked if we could retain a small amount and spend it on writing up our lessons learned from this failure. The Lemelson Foundation enthusiastically supported this suggestion. Admitting failure and documenting why you failed is not typical for donors or nonprofits. More than ten years after publishing the report on our failure in creating a new humanitarian landmine detector, I was still getting thank-you emails from technical people who were thinking about building landmine detectors and found our lessons-learned report useful.[5]

If the codevelopment process goes well and avoids hitting unexpected roadblocks, you will find yourself with an initial product that delivers some value to prospective users. You're ready to launch! Of course, we now know that these initial products are far from done. Launching a product is a down payment on years of future development to maintain and expand it. Something always needs doing in modern tech product development. This is especially true in the startup phase.

You should oversupport your early adopters. These users are investing their valuable time in trying to use your product. You should put in uneconomic levels of support into these early users because they are the best source of information about the product. Making them happy will make your product better and provide you with success stories and positive word of mouth. Of course, putting in uneconomic levels of support to the people you serve is simply nonprofit business as usual!

For projects focusing on external users, you may realize that certain types of users will not find the product helpful. This should help you better target your outreach efforts (i.e., marketing and sales activities), set accurate expectations, and reduce the amount of time wasted by both the ill-matched users and your team. Avoid promising benefits you can't deliver because doing so is an essential part of maintaining a trusting relationship with the community you are serving.

If you've made it this far, congratulate yourself. You've taken your product from an idea to real users getting daily value from your tool. For some leaders, this may be your main goal: you built the tool for your organization's internal users, and it's working. But most leaders have a product that not only touches the community served but also has a large number of potential individual users who need to learn about and, you hope, use the product. Or your expected users may be other organizations, and you want to go from twenty early-adopting organizations to two hundred or even twenty thousand. Your next step is to scale up. How can you go from serving a small number of early users to serving the higher numbers you dream of?

## THE GROWTH PHASE

Assume you now want to increase your impact one hundred times or more. Tech solutions have a natural affinity for scale, making it easier to reach far more people with your programs or your tech products. You will often be faced with the question of how to inform prospective users of the glories of your solution and how it will make their lives better. During the growth phase, most of your budget will easily go to tasks other than more technology development. You will need to do outreach, convert interested parties to active users, train them, support them, and almost certainly raise (a lot) more money. You will probably need to decide whether to expand into new geographies and adjacent fields (verticals).

The biggest enemy of tech for good is building something great and then not getting it to the people who need it. Not building but distributing the technology is usually 90 percent of the challenge. In the for-profit world, marketing and sales play a huge role in the success of a tech enterprise. For a for-profit, the goal is to scale up in order to deliver the huge profits required by your venture capital investors. For a nonprofit, the goal is to scale up in order to meet the mission goal of making a major impact on a social problem.

The nonprofit sector is typically uncomfortable with business terms such as *marketing* and *sales*, preferring terms that carry less baggage, such as *outreach*. However labeled, these functions are usually needed for the growth of the successful enterprise. Marketing is the process of communicating with potential users, informing them of the existence of your great product, and sharing why it would be so beneficial to them to use. Salespeople exist to convert prospective users into actual users, building on the work of the marketing team.

In some tech-for-good activities, the efforts to generate demand and gain users look like your existing program outreach and operations activities. The users of your tech probably think you're simply delivering counseling, technical assistance, education, health care, and advocacy, not a piece of technology. However, if your goal is to go from helping fifty thousand people a year with your program to helping five hundred thousand, you will need to invest considerable effort into finding ten times the people to use your product.

Even free products need marketing and sales. You may make a terrific free tech product, but if nobody knows about it, then it will not justify all of the effort that went into creating it. In the for-profit world and its focus on profit margins, it's often easy to throw money at marketing because there is a straightforward business case for it: spend $100,000 on advertising and ideally make $500,000 in sales as a result.

Social sector organizations rarely have significant money for outreach/marketing. Even if there is a powerful social impact case, funders generally aren't thrilled about their money going to Google or Facebook for ad campaigns. So how does the ambitious nonprofit stretch its limited outreach budget?

My advice is to lean in on your mission. When Benetech made products for special education students, we had great responses from our Bookshare email campaigns, which are an inexpensive marketing tool. We made extensive use of Google's Ad Grant program as one of its very first users. We were solving a problem that many schools and school districts had: they became the primary source of new user signups of students with disabilities. Partnerships with organizations already working with your target user communities are likely to be a very attractive path to wider distribution and social impact. Each of the tech-for-good organizations I've mentioned has solved the existential challenge of distribution in its own way and are well worth mining for ideas that might align with your situation.

### THE MATURITY PHASE

Your project is successful, and you've reached the kind of scale you dreamed of. The people you set out years ago to help are using your product, and it's working. Every successful project reaches a point where it's mature. Your cost per person served goes down as the number of people you serve goes up. Your success creates a deep well of reputation and trust, which you can use to contemplate new activities and expand your impact.

You might be wondering about adding a new tech-enabled product or program in your current field or expanding to even more geographies. You might be considering expanding into adjacent fields. You might have a partnership with the government, which would bring larger impact numbers and funding.

Running a mature organization provides you and your leadership with an exciting opportunity to make larger change beyond just your organization. You may join with others to reform your field or to create technical standards to make it easier to work together. You might have strong ideas regarding how government policies should be changed based on your experience in serving communities in need. You probably have great stories and powerful data to back up your case for policy improvements.

There are also new pressures for the successful, mature tech-powered organization. Your team might become complacent, so the pace of innovation slows. You might have a new competitor who is using a new generation of technology that threatens to displace you. Tech projects generally do not age well. The longer they have been around, the farther behind the technology is, and maintenance and improvements get harder and more expensive. One cool new feature might just break two old features that many users use daily. Every long-lived tech-for-good project is going to need to be refreshed as the years go by. This natural aging process is of course something that creates dinosaurs who have outlived their usefulness. Don't become a dinosaur!

Modern software platforms in the for-profit world are being updated, improved, and rewritten all the time, often multiple times a day. Hardware products go through regular improvements (especially those with software running inside them). The for-profit world makes enough money to pay for this amount of effort.

That updating process is harder in the nonprofit world. You might have to channel a significant amount of attention and funding into rebuilding your tech product, which is often hard to sell to your funders. The rebuilt product will need extensive testing to make sure it's reliable. Users will resist change: Why should they give up something they happily use daily? One solution to the funder and user challenges is to combine exciting new features with the (boring) rebuild, so there are new benefits to sell. The challenges of funding a needed tech overhaul in the nonprofit sector usually means that the overhaul happens less frequently than it should. You might wait five or more years to do a major refresh when you might have done it every two years if money were not a problem.

### TRANSITION AND EXIT

Even if you have committed more than ten years of effort to a field and have changed the practices in a field for the better, the day will come when your highly successful product or program is no longer at the leading edge. Given how technology evolves, it even might no longer make sense. Although I strongly encourage tech-for-good innovators to make a long-term commitment to the field, I also remind them that a tech product does not live forever! At some point, you will need to contemplate how to guide your tech and your team to a responsible exit.

I suggest you consider eventual exits even at the very beginning of your enterprise. It might come to pass that market failure is no longer an insurmountable barrier for the communities you serve, and the commercial world now meets the great majority of the social need with their products. Perhaps most groups in the field have copied your innovation, and so your unique efforts are no longer critical. Your product's functionality might have become a free feature in standard commercial products available to everybody. Should your enterprise now spin out of a nonprofit and become a for-profit company, able to access much more capital for product improvements and marketing? Does the product no longer make sense inside your organization, and might it make sense for you to spin off as an independent nonprofit organization? Or merge with another nonprofit? Each of the foregoing exits have happened to one of my past tech-for-good projects.

As a leader motivated by social impact, you will find it hard to let go of something wonderful you helped create. At the same time, the hard truth is that the greater good is often better served by shutting down your project rather than continuing to carry on as it becomes less and less relevant. Continuing a nonprofit social enterprise well beyond its "sell-by" date also poses an opportunity cost for society. Would the funding and team of the aging tech social enterprise be better deployed in something new?

I've personally gone through multiple exits associated with Benetech and several of the case studies presented in this book. The sale of the Arkenstone reading system enterprise to a for-profit more than twenty years ago provided the funding for my team to start the Bookshare and Martus social enterprises, and to acquire the Human Rights Data Analysis Group. The latter was founded originally at the American Association for the Advancement of Science and was the first group to use data to better understand large-scale human rights abuses.

Exits played out differently for these ventures. The Arkenstone products are still available commercially even though they have been long obsolete. After fifteen years, we shut down the Martus secure software for human rights defenders as donor funding declined.[6] It was also getting prohibitively expensive to maintain the security of the Martus app against attacks from repressive governments. After almost a decade at Benetech, the Human Rights Data Analysis Group spun off as an independent NGO and has since grown its programs and impact. Benetech also spun off several small and/or moderately successful products after a few years because we wanted to focus on a few larger enterprises. And in 2022 Benetech chose to focus exclusively on Bookshare and related work in accessibility for people with disabilities.

Many tech-for-good organizations founded before 2010 are now confronting how their efforts work in a changed world. Old tech infrastructure might need a complete overhaul, but the funding simply isn't forthcoming anymore. One example is Medic, which supports software solutions for the community health worker field. Medic originally started as part of FrontlineSMS, a pioneering tech-for-good nonprofit started in 2005 to use texting in social change programs. In 2010, the Medic team spun out of FrontlineSMS and is still going strong as of 2025 (but is not focused on SMS anymore). However, FrontlineSMS itself shut down in 2021, with then CEO Sean McDonald publishing a detailed series of blog posts chronicling the twists and turns of its history and his take on why it was time to close.[7]

Exits don't happen just to organizations, of course. They also happen with regularity to individual leaders. If you are one of the senior leaders of your social enterprise, it is important to develop a succession plan for your organization. You owe this decision to the mission, the community, and the team. Unplanned exits happen at every stage of a career, whether it's an urgent medical or family situation, an appointment to foundation or government leadership, or even an unexpected election to the national legislature (which happened to a nonprofit chief technology officer I knew!). A succession plan might be considered a needless time investment during the exploration phase, but by the time an effort enters into the growth phase, that plan should have been made, even if you hope it never needs to be used.

When I first started attending the Skoll World Forum more than twenty years ago, succession was one of the top topics for my peer social entrepreneurs. That's been true ever since. However, discussions of this sort tend to become reality over time. The majority of the people who won Skoll Awards in the awards' first decade are no longer CEOs of their organizations. Some have retired in the natural course of things. Some started new organizations or took over other organizations. Some have gone on to work on bigger field-level initiatives.

It is usually hard to step aside from a senior leadership position of a nonprofit. In the for-profit world, leaders often get rich when they step aside as their companies are purchased or merged. That's not the case for leaders of charities, who don't even get a gold watch! Nonprofit leaders are not doing what they do for the money. Their identity and sense of self-worth can be wrapped up in the social impact role, which makes it even harder to depart. Sometimes, though, it is in the organization's interest for a leader, even the CEO, to step aside. The most important role of a nonprofit board of directors should be the hiring and, if necessary, firing of the chief executive. Beyond the obvious reasons connected to poor performance or misconduct, the organization may have reached a phase of its existence that calls for a new kind of leader.

I wish I could share the magic formula to guarantee successful successions, but I haven't heard of one yet. However, there are steps that increase the odds of success. The first is to have an emergency plan in place to ensure continuity for your organization if you are not available for whatever reason. Try to ensure that critical information is available to your senior team and/or board members. This is especially important for money-related issues, such as requesting government funds or approving payroll. Share critical donor relationships to build more connections. Redundancy is a source of resilience for organizations.

Next, you should consider developing internal candidates to succeed you. This is harder in small organizations, but you owe it to your team to help them develop on their career path. Someone who might not be ready to take your role today might get there in three to five years. If you are the CEO of your organization, engage your board about the question of internal and/or external candidates because you might not be part of the process of choosing your successor.

Even though the idea of exit is difficult or unimaginable for you as an individual or for your organization, your responsibility as a good steward makes it important to prepare for the future with a positive legacy. Many nonprofit founders have found great opportunities after leaving their organizations, whether in starting a new nonprofit, leading a field-wide initiative, moving to the donor side, joining government as an agent of change, or even going into business!

## CONCLUSION

The entire life cycle of a successful tech-for-good effort is a useful construct as you begin to think about creating something new. There is a tendency to focus on the challenges of the initial development of a solution, but successfully completing the initial version is typically less than 10 percent of the work of delivering social impact with a tech innovation over an extended period. Taking a longer-term view and making decisions about succession and exit are the responsible actions to take given the many people depending on your innovation.

Of course, there are many important ingredients to achieving long-term impact with tech. The three most critical in a tech effort are money, people, and intellectual property—the subject of the next chapter.

# 7 FUNDING, TALENT, AND INTELLECTUAL PROPERTY

Building a technology-for-good solution presents a different set of challenges than those facing the typical for-profit tech company or nontechnical nonprofit organization. Or perhaps it's better to think of them as being similar challenges with multiple twists! Three stand out for me that affect tech-for-good efforts of all stripes: funding, talent, and intellectual property (IP). These issues are critical at all stages of the life cycle of your technology effort.

Finding funding is always a central job for any organizational leader. But in for-profit and nonprofit tech, fundraising is based on different value promises to the funder. For-profit funders are interested in private gain: how their investment will deliver exciting financial returns back to them and thus justify the risk of investing in a technology company. Nonprofit donors generally don't expect their grant to be repaid to them. They are interested in public benefits, and the exciting returns are social ones. How did people in need benefit from their generosity?

Hiring talent is also a major task for any leader, but hiring tech people is a unique skill and more expensive! Your success as a leader depends more on your team than it does on you. This makes it important to find the right people for your team.

IP plays a bigger role in technology-based nonprofits because many of their assets are intangible, not physical. The output of technology development is IP and needs to be safeguarded even while it is given away.

If you can find the money, the tech people, and manage the IP challenges, you have the foundational elements of tech-for-good success.

## FUNDING

The number one question I get from social good leaders is how to find money. Finding it is even more challenging when the request is for creating technology rather than for more clean water, food, blankets, tutoring, or whatever else people need right now. The world needs much more than money and technology, but both come in very handy when you want to create large-scale positive change.

Luckily, more and more donors are open to funding technology. Contributing to this trend are donors who care about tackling root causes of social problems rather than simply treating symptoms. Some donors are more open to taking a risk on innovative approaches. These donors are OK with the majority of their bolder bets failing because they figure these failures will be valuable learning experiences. Some donors even actively like to do tech investments because they believe this is one of the highest-leverage opportunities for philanthropy. Unsurprisingly, I heartily agree with them!

Just like a new tech enterprise moves through different phases in terms of tech and product development, as discussed in chapter 6, your funding sources in each of these phases will be different. Successful funding of a new effort requires matching up the right donors for each phase in the growth of your organization.

When you are just exploring a new idea of tech innovation, you are spending the least amount of money, relatively speaking. You are verifying what you want to build. It's hard to work full-time on a new idea when you are just talking to people about what they need. More importantly, you are trying to figure out whether your idea is worth pursuing or if you should pivot to something else.

Early exploration is often done without dedicated funding. As a social entrepreneur creating a new organization, you will be putting your time into your initial idea without expecting payment. Sometimes you will be able to get others to volunteer to help get the project off the ground. This volunteer labor is called "sweat equity" (even if the end result is creating a charity). Or a more established nonprofit leader might be exploring inside their existing organization, using unrestricted funds, or fitting the exploration under an established program's funding stream. Established organizations need to invest in their future, and it often takes looking at five or ten ideas to find the one that will become the next serious program. You sometimes can find funding that is targeted specifically for your new idea. There are often prize competitions for new leaders and new organizations. Funding in the $10,000–$25,000 range is not very much to a larger organization, but when you have zero money, these sums can be quite meaningful. Last, leaders with a track record can often get donors to give them dedicated funding to work on a new idea.

My experience is that most of the ideas I have had (or that other people have had and asked me to evaluate) did not make it through to the end of initial explorations. In chapter 6, I mentioned CityOptions, my first climate-related project idea, which had to be shelved when it did not garner enough user excitement. Unfortunately, an idea has to be very exciting to users to have the potential of justifying a major donor investment in taking the concept forward.

During the exploration and initial development phases discussed in the previous chapter, you usually will be 100 percent dependent on risk-taking donors. It's hard to convince donors to give you a grant when your idea is only that: an idea. It's just a dream with a nice mock-up! It requires a leap of faith to believe that you will deliver something amazing next year. It's also very hard to charge customers for your product when it doesn't exist or work yet.

Once you have confirmed your idea, created some prototypes to demo it, and come up with a tech roadmap for the actual product, the next stage is initial development. This is when you need to start seriously raising money because you will need it! For small or internal efforts, the needed amount could be on the order of $100,000–$250,000 or even all the way up to $1 million for larger projects. Keep in mind that the chances of failing at this phase are still fairly high, and this possibility should be communicated to potential donors while still demonstrating confidence in your team's ability to meet the challenge.

Tech-for-good donors invest in an idea to create a persistent asset that will deliver value over an extended period, especially if your tech can benefit the wider field. In the software-for-good area, many such projects are called "digital public assets" or "digital public goods." These labels imply open licensing, which I discuss more fully in the "Intellectual Property" section later in this chapter. The idea here is to build common infrastructure that could benefit many groups in a given field. If your new idea has the potential to be useful to the majority of grantees in a donor's portfolio, you should definitely get a hearing!

If you are following my recommendations for codevelopment, you will also benefit from the time investment provided by your early set of prospective users. These users have committed to testing your product and giving you feedback. For organizations that need your new product and are reasonably well funded, the chance to have a seat at the table helping chart the future of your product is compensation enough. You may be lucky and find a large customer who really needs what you're building and so provide the funds for you to do technology development, but this funding is rarely enough for your entire initial development phase.

So, where are these risk-taking donors who are willing to invest in tech?

- *Large foundations that like tech for social impact.* My initial funding for Tech Matters came from Schmidt Futures (named for Eric Schmidt, the former CEO of Google). Schmidt once had an Innovation Fellows program aimed at midcareer leaders who wanted to start something new and innovative. (I am always delighted to be described as midcareer when I think I am a little farther along than that!) Another example is the Patrick J. McGovern Foundation, which has a particular focus on tech and data innovation.
- *Fellowship programs.* In this now rich ecosystem, there are dozens of fellowship programs aimed at early-stage innovators, such as Ashoka, Draper Richards Kaplan, and Echoing Green.
- *Accelerators.* The sole role of accelerators is to support newly started nonprofits with building capacity, identifying funding opportunities, building up the network with other related organizations, and so on. The prime example of an accelerator for nonprofit tech-for-good startups is Fast Forward. Some nonprofits also go through the leading accelerators known for incubating for-profit startups, such as Y-Combinator.

- *Tech companies.* Many large tech companies offer significant grants to fund tech innovation in the social sector. Among them are Okta, Twilio, Box, Google, Microsoft, and Cisco. Many of them are "Pledge 1%" companies whose founders earmarked 1 percent of the company's stock, profits, product, and employee time to social good. Some of these companies have gone public, and 1 percent of their stock became worth many millions.
- *Intermediary organizations that operate funds for making grants based on funding from larger organizations.* For example, Tech Matters' crisis helpline project, Aselo, has benefited from grants from Safe Online and the Agency Fund, intermediaries that focus on improving online safety for children and expanding individual human agency.
- *Large donors, such as the US federal government and United Nations agencies.* Although government and international funders are not known for taking risks, over the past decade I have seen many government- or agency-funded efforts focused on innovation that matched money with risk tolerance. Even big donors are beginning to recognize the value of tech innovation to their objectives!
- *Large grants to nonprofits that are mainly not for tech but can partially (around 5 to 10 percent) be used for tech innovation.* You may be able to partner up with a large nonprofit and get funding you couldn't get for your own organization or just for technology work.
- *Individual donors.* Many business leaders are used to placing bets on people and extend this practice to backing new nonprofit leaders. They often tap into a donor-advised fund to provide these grants.

Some combination of donors should get you started. Of course, just as the tech landscape changes quickly, the donor landscape also changes quickly. By the time Tech Matters was five years old, three of our major donors had already stopped making grants because they had sunsetted (given away all their money) or the living donors had shifted their philanthropic directions.

Once the product you are building is done or close to done, you will be able to start receiving funding from more traditional donors who are either less interested in tech or less comfortable with risk taking. These donors are critical as your venture grows. You may be able to get a grant where you promise certain programmatic outcomes while de-emphasizing the fact you are using tech to get there, something that is much easier for internal tech improvements. One of your partners who is counting on you for the technology tools might subgrant you 10 or 20 percent of a larger award it has received. If your technology is working great in country X, it's a lot easier to convince a donor to fund you to expand to country Y.

Many social entrepreneurs realize that they need to engage with governments to reach systems-wide scale. Governments spend massive amounts of money delivering needed social goods. Although agency leaders and policymakers are often resistant to change, a tech innovation that helps meet a social need at much lower cost or meets it better or both can often get their attention.

**Mr. Jim Goes to Washington: Bookshare Finds Funding in an Unlikely Place**

Bookshare, Benetech's online library built by people with disabilities for each other, was an impact success but struggled with sustainability. We charged $50 a year for access to Bookshare and attracted a few thousand subscribers quickly. However, a few years after launch the Bookshare operating costs were roughly $1 million a year, but our subscription revenue covered less than a third of that budget.

Not long after the sale of Arkenstone (which created the funding to build Bookshare), we were lucky enough to attract significant funding from two key executives from eBay, Jeff Skoll and Pierre Omidyar, as well as from Jenesis, the family foundation of Ron Jensen, who was an insurance company founder. This was a major turning point, and we were lucky enough to find entrepreneurs willing to bet $1 million (or more) on Benetech as a nonprofit tech startup. The funding extended our runway for Bookshare and other Benetech projects.

Still, the question remained whether we could find a sustainability model for Bookshare. It frankly did not look like we could, and the Benetech board and I were very worried. We came up with five different plans for survival. However, it was a sixth option that saved us, and it came from an unexpected quarter: Washington, DC!

A friend, Donna McNear, was a rural teacher of children with visual impairments in Minnesota. She was also a big fan of assistive technology in general and of Bookshare in particular. Donna kept urging me to go to Washington and tell the federal Department of Education about Bookshare. "Washington!" I exclaimed. "Isn't that where good ideas go to die?"

Donna was insistent, and I soon found myself presenting Bookshare to Glinda Hill of the Office of Special Education Programs at the Department of Education. At the end of my presentation, I noted that Bookshare was at least ten times more cost effective than the two incumbent libraries for blind and dyslexic children, which I saw as the long-standing dinosaurs in the field. But that each year Congress earmarked funds for the dinosaurs, the American Printing House for the Blind and Recording for the Blind & Dyslexic (RFB&D). So I thought that even if the department liked what Bookshare offered, we would never get the chance to compete for federal funding. However, I was wrong.

Just one year after my presentation, control of Congress shifted from one party to the other, as it often does, and many earmarks did not get made as a result. The $14 million of annual funding for RFB&D was still in the budget, but Congress omitted the earmark language from the legislation. The Department of Education took half of the money to do something else with it. Glinda Hill and the team at the Office of Special Education Programs remembered my presentation about the wonders of Bookshare and e-books, and with the remaining funding they ran an open competition to provide books nationally to American students with print disabilities, such as vision impairment and dyslexia. The Bookshare proposal promised to do five times more with less than half of the $14 million of funding RFB&D had been receiving each year. As a novice bidder, Benetech won the national contract in 2007 over the incumbent nonprofit, RFB&D, promising to provide an accessible digital version of any book needed by a student with a print disability in the United States.

In one year, Bookshare went from losing money with a budget of $1 million a year to breaking even with a budget of more than $7 million a year. Bookshare had made the turn to sustainability, thanks to one giant federal five-year contract. As of 2025, Bookshare was in the middle of its fourth five-year contract with the Office of Special Education Programs. Federal funding turned out to be the key to Bookshare's sustainability.

### Breakthrough Gifts

During the early 2020s, a new class of breakthrough gifts came onto the scene. The philanthropy of MacKenzie Scott helped define this new type of funding, where an organization receives a particularly large unrestricted gift. Scott has already given away more than $19 billion in her first five years of giving. Many tech-for-good organizations have received these gifts. For example, Nexleaf Analytics received $12 million in 2022, when Nexleaf's annual budget was around $5 million.[1] Breakthrough gifts create the space for a nonprofit to step back and think big.

Believe it or not, it is even possible for a tech nonprofit to have an initial public offering. Because charities do not have shareholders, it is not possible for a nonprofit in the United States to sell shares, but it is possible for them to make a debt offering. TechSoup Global, the tech nonprofit that manages donation programs for for-profit tech companies, was able to make a direct public offering of $11.5 million, where individuals investing as little as $50 received notes committing TechSoup to repay them with interest over a five-year period.[2]

Some donors will also consider investing in a for-profit tech-for-good company that focuses on a social need. It is even possible for foundations to make grants to for-profits for this same purpose (although most shy away from doing this). Of course, for-profits can also tap regular investors who care mainly about financial return. As I noted earlier when discussing the choice to be a nonprofit or a for-profit, this investment by a for-profit may result in either mission drift or an opportunity to scale your mission.

For example, innovation in African agriculture is seen as fundable by for-profit investors, who have put more than $10 million into each of three for-profit tech-for-good companies: Babban Gona, Angaza, and Farmerline. The first two are even Skoll Award winners, an award that usually goes to nonprofit organizations.

An outstanding characteristic of tech-for-good solutions is that they can often generate revenue. The magic of technology is that although it often costs a million or two to create in the first place, sharing that tech with many people or organizations can often be done for very little. The same leverage that makes tech billionaires can be harnessed to do a huge amount of social good. When I talk about pursuing earned income, or revenue from customers using our products, many people leap to the conclusion that I am with a for-profit business. It is quite common for people to have a binary view of the world: businesses sell things for money and profits, and charities give stuff away.

However, treating your users as paying customers has several benefits. Customers have more power than beneficiaries. They expect a good product and good service. Even if "nonprofit customers" aren't paying the full cost of delivering the product and often aren't in the world of tech for good, they behave differently than someone who is receiving a gift. And that behavior is good for the quality of your product and your team because those customers expect more. Tech by itself does very little: the social impact of tech depends on how people deploy it. Paying for a product creates a sense of ownership, which leads to buy-in, which ideally leads to far more use of your product and far more positive social impact.

Earned income also gives you some degree of independence from donors. If after a couple of years your product is earning the majority of its budget from earned income, your enterprise will likely survive the loss of donor funding. The economics of technology is that once it is built, you can continue maintenance and improvements on a smaller budget. Benetech helped develop Miradi, a project management product in the environmental field, and Miradi continued for more than a decade with an annual budget of less than $250,000, much less than the initial development phase cost. I don't recommend greatly reducing your ongoing tech investment, but when funding is limited, the social impact can continue.

You also have a great deal of pricing flexibility when you have a product based on IP—intangible assets such as software, content, and data. The intangible nature of IP unlocks the powerful tool of cross-subsidization, where you can choose to provide your product to users in emerging economies for half or a tenth of what you charge in developed economies. A lower price is often enough to cover your marginal cost to support those users even if it doesn't pay for your full costs or recover the initial investment in creating the product.

An example of marginal costs coming into play is when you are operating a cloud platform. The cost to pay for an outside cloud vendor for a given customer might be $100 a month, and your customer might be willing to pay you for that even if they don't want to pay a portion of your team's current salary costs or offset the $1 million of donor funding that was needed to build the original product.

You can always choose to offer your product for free to certain users or even all users if you can keep the project going without needing earned income.

All the products I've built over the years have relied on customers and donors based in the rich parts of the world to cover more than 95 percent of the enterprise's budget. This is true even when most users were not in the most developed economies. The benefit of being a social enterprise built for social impact is that your objective becomes to help the maximum number of people while finding enough money to break even. Breaking even means being able to pay your staff to keep supporting your users and working on your product!

### TALENT

Building a great tech-for-good product requires tech people who know what they are doing. To follow the best practices I have outlined so far in the book ideally requires a team of professionals with the skills and training to fulfill different roles in the different phases of a tech enterprise. The good news is that the success of the tech industry means that there are huge numbers of such qualified professionals. The bad news is that they are used to getting paid a great deal to do their jobs in the for-profit sector.

The tech industry gets a bad reputation as populated by a bunch of self-centered people out to get rich. It's not an undeserved reputation: the industry does have many self-centered, super-rich people whose wealth is based on exploitation of one form or another. Luckily for those of us in the nonprofit sector, an increasing percentage of tech billionaires (or people connected with them) are using their wealth for philanthropy. This trend has been exemplified by those signing up for the Giving Pledge, who commit to give away at least half of their fortunes while alive.[3]

However, 99+ percent of the tech people I know are not super rich and are unlikely to become super rich. They want to work on challenging projects and make a decent living that allows them to support their families. Most of them would rather be working on the big problems of the world than on improving advertising to sell products people really don't need or a hundred other things industry does that don't help the planet. Some of them are even willing to make less salary to work on something more socially beneficial.

To staff the technology-for-good field, we don't have to change the hearts and minds of everyone in the tech industry, just 0.1 percent of them! If you can make the case that your product is going to help people and the planet, you should be able to recruit the people you need. Many technologists are hungry to help make the world a better place. Working for a nonprofit is an opportunity to go beyond just being "against bad" to actively being "for good." I call the transition "moving from money to meaning."

I run into technologists at every stage of their career, from fresh graduates to midlevel experts and managers to senior executives, and some in all these groups are eager to make the move from money to meaning. Over the past couple of decades, working for a nonprofit is no longer seen as a career-ending move, a one-way ticket out of industry. We are shifting to a world in which technologists have careers where they move more easily among various roles in big companies, for-profit startups, nonprofits, and government. Each of these experiences should make someone stronger for their next role, wherever it might be in terms of size or organization.

This fluidity and acceptance of a career path that spans public service and private practice have been the norm in the legal field, especially in the United States. In 2013, the infamous Healthcare.gov fiasco, where a handful of Silicon Valley tech types rode to the rescue of an important new government website that didn't work, led to much greater interest from tech professionals to go into the federal and state governments to use technology to improve public services. Following this incident, the Ford Foundation began working to support a public interest technology movement, hoping to replicate its success in pioneering the public interest law field in the 1960s. In addition to getting more technologists to work for social good in the nonprofit and government sectors, the foundation is also actively trying to bring a more diverse workforce into public interest technology and the tech industry in general. Tara Dawson McGuinness and Hana Schank recently wrote an excellent book about the public interest technology movement, *Power to the Public* (2021), with a particular focus on tech for good in government.[4]

The people you will need for your tech-for-good effort are out there. It just takes longer to find them. Tech Matters is always looking for someone who wants more than just "a job." In our experience, alignment with our social good mission is critical to building a strong team. We look for evidence of a heart in our applicants. Seriously. Have they ever worked for a nonprofit or volunteered for a cause? If there is nothing in their past that indicates an interest in their fellow human beings and/or the planet, they are unlikely to be happy or successful working for a nonprofit that doesn't pay the wages supplied by the big tech companies such as Google, Apple, Meta, and Amazon.

Compensation is a huge topic in recruiting talent for tech-for-good organizations. Because most of these organizations are organized as non-profits, they rarely can match the compensation paid by industry. Potential employees often take a pay cut to work in tech for good, and the more senior they are, the larger the pay cut tends to be.

Tech Matters is transparent about its pay scale in its job announcements, as are many of its peer organizations. Everyone saves a lot of time if a possible candidate is not willing to work for the amount the nonprofit has budgeted to pay and so chooses not to make themselves a candidate for the job. We try to bring up salary in the very first interview as we assess the candidate's social motivation (Are they passionate about social change?).

Successful organizations developing tech for good approach these challenges in similar ways. Tech Matters initially hired only people who live in California, but the pandemic caused us to reconsider this restriction as working in an office was impractical for a couple of years. We shifted to a virtual model, and Tech Matters now has a three-pronged approach to talent:

- *Compressed wage scale.* We try to pay our frontline team members salaries that are comparable to what they might get in the for-profit world (in their country). Our more senior people are likely taking a big pay cut compared to for-profit norms.
- *Hire globally, pay locally.* We recruit talent internationally, which helps in terms of supporting our users and having team members closer to our partners, while paying based on location.
- *Full team member status.* No matter where a team member is located, no matter their legal status as an employee or a contractor, we work to build a team of equals. As a small organization, we try to control the time zone differences among team members to ensure they overlap in working hours in order to encourage team interaction.

This approach helped us grow from four people to twenty-five people during the pandemic, and we think it's working well. We have talked with our peer nonprofit organizations, many of whom have taken similar approaches in terms of pay structure and virtual workplaces.

However, many tech-for-good organizations have stuck with building a team in a physical office. This is especially beneficial for team members earlier in their career, where access to more experienced colleagues provides access to more mentoring and personal development.

Finding the right people for your social enterprise is crucial. A compensation strategy is just one piece of the puzzle. You must determine the skills and experiences you need for different roles during each phase of your social enterprise's life cycle, and then you must find the people who have them.

The following roles are quite typical in tech-for-good social enterprises, just as they are in for-profit tech companies. First, you need someone to lead this effort. The leader needs to be able to inspire trust from other people, whether funders, team members, or partners.

Ideally, your founding team will soon have two key leaders, one in product management and the other in technology development. The product manager is responsible for representing the voice of the users. This person typically conducts interviews and user testing as part of developing the product. The tech leader is responsible for recruiting engineers and developers to build the product, all in close partnership with the product manager. Finding great product and engineering managers is hard, but it is even more difficult to try to deliver a product with leaders who haven't done it before. I look for five to fifteen years of experience for both roles, generally acquired in the for-profit tech industry.

Of course, other important roles are common to most social good organizations, such as administration, human resources, accounting and finance, and fundraising (sometimes labeled "business development"). Your new tech effort may have the benefit of a larger parent organization or a fiscal sponsor (a common structure for new nonprofits in the United States) to take care of most of these roles. These functions are often outsourced to external vendors.

Once your product is ready to launch and begin serving users, you will need to fill additional roles. These operational roles include outreach, training, customer support, and technical support. You might just have one individual doing multiple jobs because you can't afford to hire different people for each of them. A stand-alone product organization with aspirations for impact over the long term should probably be aiming to have ten to twenty team members across these different functions, which I think of as the minimum viable team to sustain a tech-for-good social enterprise, even if some amazing teams are successful with fewer members. Of course, some organizations grow past a staff of one hundred if they have the demand for that amount of support from their partners (and the funding!). However, an internal tech team supporting a larger program-driven organization is typically staffed by just a handful of people. The great thing about tech for good is that it can be possible for a modest-sized team to build a product used by millions.

The question of volunteers often comes up, and it's a delicate one. Tech people are well paid, and it's tempting to use part-time volunteers instead of paid team members to save money. I find this to be a false economy. Part-time volunteers usually have other, higher priorities in their lives, which will come before their volunteer commitments. It's hard to be a director of engineering for only five hours a week. Things will go much more slowly, and turnover in volunteers working in key roles can set back your project, especially if the new volunteer thinks everything done by the previous volunteer needs redoing! This happens with alarming regularity among software developers.

However, there are several great uses of volunteers in tech-for-good organizations. Before the organization has any funding, the founders are typically volunteers. As the organization evolves, it may be that someone who very much wants to work full-time for you will volunteer while waiting for the funding to come in to allow them to quit their day job and commit all their time to your organization. Volunteers are also terrific for filling in expertise you can't afford but very much need. It may be that your core team members, who are frequently generalists, need one of the world's top experts in a specific field to spend two or three hours advising them. Those three hours might save your team a month (or more) of floundering. Impossibly expensive consultants often choose to provide their help pro bono for a short period if they are asked nicely with an emphasis on the social impact opportunity. In addition, all my social enterprises have made extensive use of pro bono legal assistance. Only a large organization can afford an in-house attorney, even on a part-time basis.

You need to remember that volunteers are not actually free. The social bargain with a volunteer is to respect their valuable time and give them a good volunteer experience that gets them to feel as if they are making a difference. Tech volunteers do not want to be doing the technical equivalent of stuffing fundraising appeal letters into envelopes.

On the cost side, it takes time away from your team's work to prepare a volunteer project and work with volunteers to deliver them a good experience. At Tech Matters, we often turn away great volunteers because we'd rather our team focus on their top priorities without distraction. This is an even larger issue if the volunteers are less skilled than the existing team. Larger, more mature organizations tend to have more capacity to engage student interns and are more willing to see the opportunity primarily as a learning experience for the interns. Unpaid internships for students are in general being discouraged, though, because this practice tends to disadvantage students who do not come from wealthy backgrounds.

Some organizations invest the time to create a volunteer program and accept volunteers at scale. A well-organized program can do a better job of utilizing volunteer time and reducing the per volunteer cost to the organization by spreading volunteer management costs over a large number of volunteers rather than creating a custom experience for each volunteer. An example I often give is the way the American Youth Soccer Organization, or AYSO, the largest youth sports program in the United States, runs largely on volunteer labor. AYSO has less than thirty employees but uses more than one hundred thousand volunteers per year.[5] Each volunteer role is highly scripted, ranging from large numbers of volunteers who serve as coaches and referees to smaller numbers of volunteers who take on administrative and leadership roles. These roles come with required training programs and detailed job descriptions, which ensure the safety of the children as well as enhanced protection from liability for the volunteers involved if they follow their training. However, these kinds of volunteer programs are difficult to scale up successfully in the tech-for-good field.

No discussion of volunteerism and tech for social good would be complete without mentioning hackathons, which come in bad and good flavors. Originally, the hackathons told a story of amazing impact. Spend a weekend and cure global hunger. Unfortunately, global hunger is quite resistant to being solved in a weekend. The volunteers were often crushed to learn that their hackathon-winning app wasn't being used six months later.

The for-profit sector depends on serious enterprises delivering products built by professionals over a period of months and years. I was often puzzled why volunteers from industry thought it was OK to whip something up quickly and hand it off to the nonprofit sector. For some companies, the hackathon was less about solving a problem for the nonprofit and more about building skills and camaraderie among the for-profit employees who were volunteered for the hackathon. On occasion, corporate volunteers never even showed up because their bosses told them to work on a priority corporate project instead of the social good project. The cost to the nonprofit to prepare for the hackathon was often not matched by the value provided.

Cheerful hackathon cynicism aside, there are terrific models for hackathons where the value to the nonprofit far exceeds the time invested by the charity's team to prepare for the hackathon. A great example is a small nonprofit tech team that doesn't have someone with user experience design skills. A hackathon where a group of designers creates several different designs for your new product can be immensely valuable if you take the best designs to your potential users and get their feedback. In addition, it may be a great outcome from a hackathon if for-profit tech people get exposed to how cool your projects are. Maybe they will become a steady volunteer, a donor, or a board member or even join your team as an employee. Plus, more people might get exposed to the entire field of tech for good, which will help organizations other than your own. However, these hackathon outcomes are pretty soft and may not justify the time and energy your team will need to invest into running a hackathon.

Volunteers, carefully chosen and managed, can deliver significant value to your project, but they rarely substitute for a capable, dedicated team that you employ.

Beyond determining your pay strategy and figuring out the roles you need to fill, the last major challenge with respect to talent is finding the critical members of your enterprise's team. It's one of the most important tasks of a senior organizational manager: to recruit a diverse team with the skills you need to build a great product.

As I noted earlier, we don't need to convince all tech professionals to come work on tech focused on social good rather than on profits: less than 0.1 percent of them will do. But it is a challenge to find that one-in-a-thousand tech professional who feels a deep connection to social change. You will need an intentional strategy to find these candidates (or "source" them, in recruiting lingo) and fill your pipeline with worthwhile people. There are many good strategies: I try to make a point of learning what our peers are doing for new ideas in recruiting. I will share the current practices used by Tech Matters as just one example.

We include a salary range for all of our position listings: this is currently considered a best practice in recruiting as part of a set of strategies to encourage pay equity for women and other historically underrepresented groups. It also is now a legal requirement in multiple jurisdictions, including California, where Tech Matters is based (at least, in theory, since we are a virtual organization). It's a waste of time for you and the candidate to spend thirty minutes speaking together when they think they need a salary 50 percent higher than the top end of the range you are offering, so that range must be clearly posted.

We list our required qualifications, but we make sure to mention that we will look at candidates who do not meet our minimum listed requirements. Research has shown that women and more diverse candidates are, compared to white men, much less likely to apply for roles when they fail to have all the listed qualifications. Since we do not always get candidates who meet all our requirements, we consider candidates who are exciting and might be able to do the job well. But they need to be encouraged to apply. Some of our hires didn't meet the minimum requirements in our job listings but still turned out to be great team members.

Even though it's considered old-fashioned by some, we require a cover letter for candidates applying to Tech Matters. We are trying to find out if their cover letter demonstrates an awareness that they are applying to a nonprofit for a job. A powerful cover letter that conveys a passion for social justice ensures a serious review of a candidate, even if the résumé wouldn't have assured that. Plus, it is an indicator of whether the candidate pays attention to directions! A great résumé with no indication that they know what we do or with only a generic cover letter will almost always get rejected.

Beyond the posting process, we make extra efforts to find more diverse candidates. We might choose to interview candidates from a South African job-listing website for technical roles where being based in Africa would be an asset. In the United States, we have experimented with a number of specialty websites designed to target women and other people with backgrounds typically underrepresented in the tech industry. We keep using the sites that brought us interesting candidates, of course.

Modern recruiting technology has made it easy for hundreds of people to apply for the job you have listed once it hits the major recruiting sites. On the employer side, there are also great technology tools for rapidly assessing and responding to candidates. The number one reason for a quick rejection at Tech Matters: no cover letter. The number two reason right now is probably a cover letter obviously written by AI rather than by the applicant!

The rest of our process should be familiar to anyone who has tried to hire team members. We do screening interviews (Do they have a heart for the work? Are we confident that they would accept an offer in the listed salary range?) and then more intensive team interviews of the finalists. It's not uncommon to be interviewing finalists who live both inside and outside the United States, where Tech Matters is based.

One last recruiting idea is our fellowship program. During times of upheaval and layoffs in big tech, my tech nonprofits have offered a nine- to twelve-month fellowship for people transitioning away from the for-profit tech industry. The pay is modest, but it includes health and other benefits. At the end of their fellowship period, roughly a third of fellows end up joining our nonprofit at our normal wages. That one-year period was often enough for us to find longer-term funding for that person. Another third of our fellows goes on to join other nonprofit tech organizations, and their fellowship with Tech Matters made their application stronger. And last, the remaining fellows go back to the regular tech industry, but we've infected them with the tech-for-good meme!

There is, of course, much more to the incredibly important work of recruiting, hiring, onboarding, and ongoing management of your team that I don't cover here. I have tried to highlight some of the key differences that set a mission-based tech enterprise apart in terms of its approach to finding the right people.

### INTELLECTUAL PROPERTY

The tech-driven enterprise is built on intellectual property. Even if you choose to give everything you create away to the world, you will need to pay attention to IP law to ensure your intent is followed.

Every tech product is built on top of IP belonging to others. Your team will almost certainly also be creating your own IP along the way. Your users might generate still more IP based on how they use your product. Whether you like the idea of intangibles being considered property, your tech-for-good venture will need to engage in the world of IP.

Depending on your project, different types of IP will be relevant. Software or content (videos, educational courses, white papers) will be in the world of copyright. The data they use or generate will usually touch on the rights of others (for example, privacy rights in personal data). Hardware products and pharmaceuticals might need to navigate patent issues. In addition, the social good world runs on trust, so your organization's name and its products' names will be important as trademarks.

Some attorneys specialize in all of these types of IP. In addition, IP laws vary by country. You will definitely need to get help in your organization's home country. Organizations working internationally will probably need to get help with the laws of other countries. You can often get pro bono assistance from law firms who want to volunteer to help nonprofit groups. I've often gotten free legal help from firms around the world who were referred to me by the Lex Mundi Pro Bono Foundation, which focuses on helping social entrepreneurs.

Getting help from attorneys in the commercial tech industry is a cultural challenge. The tech industry's approach to IP is proprietary. Businesses in the industry use IP law to lock down their ownership of IP so they can make a lot of money and stop competitors from using their stuff. Nonprofit tech-for-good organizations are much less proprietary because of our social mission. We often want to share what we create with other groups under open licenses (more on that topic shortly!). You need to find an IP lawyer who understands and is comfortable with organizations that do not want proprietary control over IP but instead want to share it for greater social impact.

Because creating IP is expensive, nonprofits need to stretch their limited funding. First, they should try to borrow critical IP for free. My own experience is that when I ask a for-profit company for a free license to use their IP for a social good, they say "yes" 80 percent of the time. This is free access to their crown jewels!

This quick agreement surprises many people because they think that for-profit tech people are greedy, controlling, and selfish. However, my experience is that most leaders are quite proud of the tech created by them and their teams. Although my investors vetoed a product to help blind people, as I shared in the preface, they liked the idea of their technology helping a disadvantaged group as long as it didn't hurt the company.

Business leaders and investors are also realistic about business opportunities and understand well why many of these social good applications do not make great for-profit businesses. So when I make a strong commitment to use their technology for social good in markets they have already given up on and to stay out of their lucrative markets, they are usually willing to make that deal. The only thing I give them in return is the promise to tell their employees and customers about the great things we're doing with their inventions, their IP. Stories in exchange for valuable IP: that's a great social bargain!

Beyond convincing proprietary owners of IP to share their tech with your social enterprise, you can also get valuable assets for free in other places. For example, I wrote in chapter 5 how MapBiomas used AI to understand how land use evolves both over long periods as well as in real time. The AI was made possible by the availability of free satellite data released to the public by the US government. The US taxpayer had effectively paid for the Landsat satellites, and the government decided to make it freely available. MapBiomas has used that IP to help people in more than a dozen countries!

When we at Tech Matters have a need to build software to meet a specific need, we try to check to see if some other group has already written software that meets that need or comes close. That software is often available under an open-source license, which means we are free to use it in our project, subject to some minor restrictions we are likely to find acceptable. We can even jump in and add an extra feature we need that the existing open-source software lacks. Open licensing is a very important concept in IP for social good, so it's worth expanding upon.

Open source uses IP law to flip the script for software by sharing it instead of keeping it under lock and key. There are no license fees for open-source software. The software source code is publicly available to all. Other software developers can look at the source code and make changes, either to fix bugs, add features, or customize the software to their needs.

There are no restrictions on who can use an open-source software program. You can use it in a business or in a human rights organization. You can be against the current government or for it. The *open* in *open source* is about freedom.

The values of the open-source approach are quite compatible with the social change sector. Open-source software is transparent (at least to someone who can read software). You can see what the software is doing and be confident it is trustworthy. You have the freedom to take the software, change it to your liking, and run it on your own computers, all while maintaining control of your own data.

Open-source software also benefits the nonprofit sector when it reduces the cost to use good software. As I discussed at length in chapter 3, Kobo Toolbox is a great piece of open-source software that makes it very easy to design a survey, conduct a survey, and analyze the responses. The software is open source, and Kobo will even host your data for you without ever looking at them because they're your data.

Kobo is a great example of how a valuable asset can be reused for more social benefit. When our team at Tech Matters was creating Terraso, a platform designed for local leaders to help them build more sustainable economies in the face of the climate crisis, we thought about Kobo. Following the design processes outlined earlier in the book, we asked many local leaders about the tech tools they needed. High on the list was the ability to run surveys in their region, often including areas with poor internet connectivity, which brought Kobo to mind. Rather than building a survey tool from scratch, we integrated Kobo into Terraso and added an extra feature we thought would make it better. All because of the Kobo team's generosity, expressed by their open-source licensing decisions, we were able to meet our users' needs with the investment of a month of developer time instead of trying to re-create what Kobo had already done well with years of developer effort.

Another great example of what open-source licensing makes possible and even encourages is what Partners in Health–Malawi did with the Medic community health worker open-source software project, the Community Health Toolkit. The toolkit is designed to make it easy to build software applications for health workers that are adapted to the local context. As I mentioned in chapter 3, Partners in Health is an extremely well-regarded community health worker nonprofit, and its late founder, Paul Farmer, was widely celebrated as a leading social entrepreneur (even though he disliked the label rather intensely). The Malawi team built a more extensive app, YendaNafe (locally interpreted as "Walk with Us"), on top of the Community Health Toolkit and then released their expanded app under an open-source license, saying that they wanted to carry on Paul Farmer's "legacy and vision . . . to bring the benefits of modern technology and social medicine to all."[6]

If you are building a long-lived enterprise, as I hope you are, open-source licensing also makes it easier to keep improving your technology. For example, I mentioned Aselo earlier in the book, a platform for crisis response helplines around the world. Each time Tech Matters does a new project, funded by either a donor or a helpline, new features are almost always needed for that project. When we build those new features, they are released as open source. This means that every helpline already using Aselo or adopting it in the future gets the use of those new features for free. I'm very proud of the willingness of better-resourced helplines to fund new features and then share those features with the world rather than keeping those features to themselves, which would be typical for most businesses. That's the ethics of being a nonprofit organization as it applies to software development!

I personally believe that trust, not money, is the currency of the social good sector. It is essential for tech-for-good organizations to work hard to be worthy of the trust placed in us by our peer organizations and, above all, by the people we serve. Open source is one element of generating that trust. If you are building your project as open-source software, you are making a declaration of joint ownership with society. It's a statement of hope: hope that your project is useful to the community. It's also a statement of trust: you are willing to create something of value, give it away without strings, and share control over it.

The value of open licensing and the trust that goes with it are not limited to software. The organization Creative Commons (CC) does the same for copyrighted content, such as essays, white papers, pictures, videos, music, and more. If you produce a public service booklet or video, and your goal is to reach the maximum number of people, you should put a CC license on it. That way, anyone who sees your content knows they are encouraged to send it to as many other people as possible. There are six different CC licenses, all of which allow free sharing but include additional requirements, such as acknowledging the original creator or mandating that the work be shared unchanged. In addition, there is the CC0 declaration, where you can put your work in the public domain without any other requirements at all.

Putting a CC license on something you have created says to the entire world that you want them to have it. The science fiction writer Cory Doctorow explains it so well: "As a writer, my problem is not piracy, it's obscurity, and CC licenses turn my books into dandelion seeds, able to blow in the wind and find every crack in every sidewalk, sprouting up in unexpected places."[7] Social good organizations are not trying to make money, and they are not trying to stop people from "pirating" their work. They want more people to use and thus to benefit from their work.

The topic of data is complex because it often involves sensitive data belonging to other people. I expand on innovative ways to manage this confidential data in the next chapter. However, it's worth pointing out here that some databases with data that can be responsibly shared are often released as "open data." I discussed earlier in this book the example of MapBiomas using satellite data released publicly by the US government to analyze land use in Brazil over the past thirty years. Some datasets are released into the public domain, and many databases are released under CC licenses, even though those licenses weren't originally designed to apply to data. Yet again, the power of open access can strongly support social impact work.

One large benefit of standardized licenses for software and content is that you don't need an attorney. These licenses were designed with help from lawyers to ensure they work correctly and then made available to anyone to use. Everyone who writes software actively knows what an open-source license means because the social bargain is easy to understand in everyday language, as in "share and share alike." Authors, photographers, videographers, musicians, and artists can easily understand what a CC license permits because the licenses' developers have taken the time to explain these licenses in plain language and simple icons. The tools of IP law were invented to stop the sharing of IP, but open licenses make it easy to free your IP to be shared with the world.

Patents are another type of IP. Pharmaceutical companies and big agribusiness often use patents to control access to IP for profit maximization. Obtaining a patent is costly, so the incentive to use patented knowledge to make proprietary products or collect licensing revenues is significant. However, the benefits of patented inventions fail to reach all of humanity, which leads to different efforts to overcome this failure. For example, the Patent Lens social enterprise hosts a global database of patents so that anyone in the world can learn about the great majority of awarded patents. Perhaps an exciting invention is patented in the rich part of the world but not in your country, so you are free to use it. Maybe the patent has expired. Some forward-thinking companies have also been granting free licenses to their patents for environmental or other social good work.

I also encourage you to consider actively asserting your right under existing IP regulations and frameworks to claim copyrights in your work and police your trademarks. You might even choose to file a patent on a new invention your organization has created before somebody else files one and then tries to stop you from using your invention. By claiming ownership of your assets, you will be able to use the existing IP system to defend your ability to use those assets and, if you choose, to give them away to others.

If your goal is positive systems change, freely sharing intangible assets should be your default approach.

## CONCLUSION

Starting up a tech-for-good enterprise is a major personal and professional challenge. Donors tend not to think about technology as they pursue philanthropy. They tend to focus on the needs of people who are suffering rather than on how tech tools might be assisting those people in need. Technologists rarely think about going to work for a nonprofit when they choose the difficult path of pursuing a technical education. And the legal profession is trained on protecting and defending IP assets, not on assisting in the most effective way to freely share them. However, each of these groups has a natural interest in solving society's biggest problems. Your job as the leader of a technology social enterprise requires you to make the connections to philanthropists, technologists, and attorneys and to gain their indispensable support for your social mission.

With adequate funding, talented people, and an enlightened approach to intellectual property, you will have most of the right ingredients to realize your vision of tech benefiting the world!

# 8 THE FUTURE OF TECH FOR SOCIAL IMPACT

Our tech-for-good journey now looks to the future. The preceding chapters took you on a tour of the importance of tech for good, the top-four bad ideas to avoid, the top-three good ideas to follow, how to design your solutions using modern approaches, and the power and potential pitfalls of AI. They also described the life cycle of a tech-for-good social project and the biggest challenges associated with starting up and running one: finding the needed funding and talent and developing IP strategies. As you consider applying technology to solve social problems, what else should a forward-looking leader keep in mind?

We are lucky that much of the tech needed for social good in the future is here today, having been developed and proven by the for-profit sector. However, for-profit businesses too often apply technology for maximum profit extraction and in ways that favor the 10 percent wealthiest, while taking power away from the rest of humanity. Thus, as tech-for-good leaders we must take the best ideas from tech advances and apply them with values such as equity and social justice and with a vision of helping all humans have better lives on a healthy planet.

As you develop your particular contribution to this vision, here are my top three recommendations:

First, we need to shift power over tech and data away from major corporations and national governments to individuals and communities.

Second, we need to embrace large-scale ethical data collection as an essential tool for solving social problems at scale and provide examples of how data can and should be used for benefit and not just for profit.

Third, AI will ultimately have a huge positive impact on humanity if we work together to collect large and representative datasets and to build common tech infrastructure to lower the barriers to using AI effectively and responsibly.

Shifting power is the hardest of these goals to achieve. I will go more deeply into this first recommendation as a result, but of course these three suggestions have considerable overlap.

## RECOMMENDATION 1: SHIFT POWER TO INDIVIDUALS AND COMMUNITIES

As the world has shifted from an industrial age to a connected information age, we have come to see the limitations of industrial-era approaches to social programs. Top-down approaches make assumptions about what people need that are rarely true. One size doesn't fit all. Government programs have their limitations. Capitalism leaves too many people behind.

As the failings of traditional approaches have become better understood, there is also widespread recognition that shifting power to the grassroots is a much more effective way to solve social problems. Most poor people in the world have the capacity to build a better life for themselves and their families but lack access to the tools they need to do so: better health care, education, training, credit, weather forecasts, and agricultural inputs; secure land tenure; and, of course, cash.

This view is increasingly shared by social good leaders. More social entrepreneurs are forming partnerships with the communities they serve. Powerful initiatives have been formed around educating girls, improving the incomes of individual farmers, making cash transfers, and expanding financial inclusion. It's no surprise that in much of the world microcredit programs are delivered by "self-help" groups, which are circles of local peers trying to use microcredit to help individuals grow their personal businesses. The social leaders and organizations behind these efforts share a belief that building up the social, intellectual, and financial assets of the communities they serve will lead to better human well-being.

Mauricio Lim Miller, the founder of the Family Independence Initiative and a MacArthur Fellow, believes in this approach so strongly that he famously made a policy that any employee who told a person in poverty what they should do would be fired.[1] His experience is that most people who manage to get out of poverty do so not because of helpful social programs but in spite of them. He also believes that funding that goes to most social programs should be rechanneled into scholarships, investments, and/or loans to support what families themselves are prioritizing.

Sad to say, though, it is far easier to talk about shifting power than actually to shift power. Not surprisingly, major corporations and national governments want to hold on to their power. But the social sector can also be resistant to change. Peaceful power shifting requires trust and the letting go of control. The social sector often has a hard time with trust and letting go. It could be donor requirements, such as those stemming from corruption concerns, that make us hold on. Or it could be that we want to incentivize better performance, but we do it in ways that disempower and discourage people from delivering the desired results.

The good news is that modern technology makes power shifting more possible. There are at least three ways you can enable it: (1) you can shift the focus of data collection to ensure it is more directly useful and relevant to the people you serve; (2) you can measure your impact and the satisfaction of the people you serve with your products or services in near real time; and (3) you can support mass customization, which is the process of designing products and services that can be personalized to meet the needs of individuals while still being delivered at a relatively low cost.

### Empower Individuals and Communities with Their Own Data

Today's data-use status quo is not a story of individual empowerment. In her book *The Age of Surveillance Capitalism* (2019), the Harvard Business School professor Shoshana Zuboff explains how the opposite has occurred.[2] Large-scale data collection enabled the rise of multi-billion-dollar companies such as Google and Meta (Facebook), each of which holds the data of billions of people. They use—and share—your data to generate immense wealth for themselves and their shareholders. Their version of "sharing" is to sell your information to advertisers (for a lot of money) so that the advertisers can sell you (a lot of) stuff. In contrast, our job as tech-for-good leaders is to use the data to help the communities we serve and to enable them to better help themselves rather than to extract ever more money from them.

The Callisto case study in chapter 3 is a good example of using data as a source of power for a group of people who have experienced sexual violence. Callisto hopes to be a source of empowerment for survivors by providing them with support and knowledge that might enable them, if they choose, to seek justice.

In 2022, I coauthored an article on decolonizing data with Nithya Ramanathan, a cofounder and the CEO of Nexleaf Analytics, a public health tech social enterprise I mentioned earlier.[3] In the article, we discuss the problem in the social sector, where data are often used in ways reminiscent of colonial appropriations of natural resources. Nithya and I wrote this article after we gave a talk about this issue at the Skoll World Forum, and many of our peer social entrepreneurs realized that they were inadvertently using data in ways that might be seen as colonial. That is, they were extracting data from communities, not sharing the data back, and sometimes using the data in ways that hurt those from whom the data were collected.

In the article, Nithya mentions a case in which a foreign researcher had traveled to an African country and conducted a study there without the awareness of that country's national health ministry. The first time the ministry heard about the study was when one of the ministry's leaders was attending an international conference where the foreign researcher presented the results criticizing the ministry's work. The country's health ministry didn't have the chance to use the data to learn and improve. Not an empowering experience, to be sure.

My point is that data are often collected, analyzed, reported, and acted upon without a clear purpose of being helpful and relevant to the people who are the subjects of the data collection. It's not uncommon for data about one place to be analyzed and used in a different place, where the people working with the data lack knowledge of the data's original context—knowledge that could or should influence the data's interpretation. For example, was the health ministry's performance down last year because of lackluster program management or a typhoon that devastated the country?

Even more concerning, data are also often used to punish people and organizations instead of to improve performance. "Paying for results" is a popular approach in modern philanthropy with the admirable goal of encouraging better outcomes. However, care needs to be taken to ensure that these kinds of program performance incentives are implemented in ways that are constructive and do not accidentally damage the intended beneficiaries. Imagine if health funding were reduced in a particular province because the province's numbers failed to meet the desired goals simply because of a natural disaster. Context matters!

This is a reminder that nothing makes data less accurate and less useful than when they're used to incentivize or punish people. Simplistic approaches that make it straightforward to provide data that supports the financial outcomes of the people supplying or collecting the data are not helpful.

For example, I was advising Gavi, the Vaccine Alliance (formerly the Global Alliance for Vaccines and Immunization) about its data challenges, and I learned that certain districts had data showing that 120–150 percent of the children in those districts had been vaccinated with the basic set of required shots, a clear impossibility. Were children being vaccinated more than once? Independent on-the-ground surveys turned up significant numbers of unvaccinated kids. In this case, financial rewards for recording vaccinations created incentives to overstate the number of vaccinations. These distorted data were not useful for assessing or improving true vaccination rates, which was Gavi's goal.

One of the top ways to ensure that data are helpful and relevant is to link their collection to direct benefits for the individuals and communities. If the data subjects see how data collection leads to more money in their pocket or to helpful and tailored advice, they will be more willing to share their data. There will also be an incentive to provide accurate data because sharing incorrect data will usually produce bad advice and useless insights. I call this approach "providing news they can use."

Well-meaning leaders often don't see how tech and data collection projects can have unintended consequences. Although my main intention in this book is to share best practices and tell relevant stories to help you avoid bad outcomes, I also think it's important to keep in mind that shifting power also means shifting the power to make mistakes. It's quite possible that individuals and organizations might not make the same decisions you would if you were in their shoes. Sometimes you just have to trust that overall human well-being is strongly enhanced by devolving power.

### Measure Impact in Near Real Time

In past eras, gathering and analyzing data was expensive, cumbersome, and slow. Countries typically performed censuses with a frequency measured in decades. Organizations collected and reported financial, program, and other performance data on an annual basis. Measuring impact rather than outputs and activities was practically impossible for the average program.

If you were lucky, a dedicated research effort might show that a program was effective with a small population. But then these programs would be scaled up by people who assumed they would have the same effectiveness on a large population. Outputs (such as the number of people reached by the scaled-up program) would serve as a proxy for impact. If the impacts of the scaled-up program were measured, though, they often frequently fell well short of expectations. This was (and still is) a widely recognized problem in the health field, where an evidence-based practice that worked in a small, controlled trial didn't work well when it was scaled up. The most common reason was "lack of fidelity," where the scaled-up program ended up being delivered differently.

We are lucky that times have changed. The cost to measure impact in a timely way has gone way down for two reasons. First, you can now use technology to deliver a program. With technology, program performance data can be collected as a by-product of program delivery more or less for free! The data allow you to "listen" to the community you are trying to serve simply by observing their digital actions. For example, if you are delivering a course to build capacity, and fewer than 5 percent of those who start the course complete it, your users have spoken to you and delivered the message: "This course does not work for me!"

One of the centerpiece programs of SameSame, an African social enterprise that uses technology to support LGBTQI+ (SameSame's preferred acronym) young people in South Africa and other countries, is an eight-lesson cognitive behavioral therapy course delivered via a WhatsApp chatbot. The course is designed to improve the mental health of queer youth. Because the entire program is provided digitally, SameSame uses the data it collects to test different user interface elements to see which ones are most effective in encouraging completion.[4] Further, SameSame was able to do a major research study that showed the beneficial impacts of completing the course on teen mental health.

The second reason the cost of measuring impact is much lower now is that technology makes it possible for you to easily ask questions of *all* the people you serve digitally, observe their behavior, and hear them implicitly. Just like businesses now tack on an optional survey at the end of a customer service interaction, you can tack on a similar survey at the end of a social services interaction. In addition, the people a social enterprise serves often will respond to surveys at a much higher rate than customers of for-profit businesses. Of course, to truly make surveying a tool for shifting power to the individual, we must pay attention to the responses.

One effective approach to increasing response rates is to include a message at the top of such a survey stating that it's completely optional and anonymous but that because the nonprofit depends on donor funding to provide the free service, it would be a big help if users of the service or product could answer a few short questions.

I've seen more and more nonprofits that use technology employ these kinds of mechanisms to listen to their users and spot problems right away rather than a year from now! Some of these technologies even allow you to measure impact in near real time—a major win in the nonprofit field of monitoring, evaluating, and learning. One great example I saw was a confidential survey question added at the end of a crisis interaction by Kids Help Phone of Canada: "If you hadn't called our helpline today, what would you have done instead?" Roughly 10 percent of the young people answered that they would have gone to the emergency room or their primary care doctor. Canada Health Infoway, the government body charged with independently measuring the benefits of health interventions, assessed that the helpline delivered much more in savings compared to emergency room or doctor visits.[5] If a conversation helps young people avoid self-harm and/or an expensive visit to a medical facility, the benefits to both them and the medical system are obvious.

### Adapt to Individuals' Needs with Mass Customization

Mass customization is a true antidote to traditional one-size-fits-all social programs, especially in areas such as education and agriculture. It turns out that every kid learns differently and every farm is unique. In such areas, a technology-supported program that gives each user what they need based on data about their specific situation and preferences could be a great improvement over a standardized offering. The affordable availability of data in digitally powered programs makes these new approaches feasible and makes the benefits of data collection more apparent to each individual.

One tech-for-good NGO I particularly admire is Precision Development, which has figured out how to deliver agricultural extension services via technology to five million farmers in some of the poorest countries of the world, many of whom will never otherwise have access to a human agricultural extension adviser.[6]

#### Bookshare: Mass Customization in Accessibility

One early example of a mass-customization approach was Benetech's Bookshare e-book library for people with print-related disabilities, such as blindness and dyslexia. Before Bookshare, the status quo was audiobooks on cassette tapes or enlarged-print or braille books delivered by mail. Braille and large-print books are physically much larger than standard print books. For example, one *Harry Potter* book in braille format weighs twelve pounds and is more than one foot thick. As a result, because of the considerable expense of converting and producing these formats, far fewer accessible titles were available to people with disabilities that affected reading, compared to the number of regular print books available to readers without a disability.

Specific physically accessible titles did not work for all readers. Braille books worked only with the subset of blind people who had learned to read braille. A reader with dyslexia probably benefited from an audiobook, as did most readers with a visual impairment. A low-vision reader may have preferred a large-print edition, but there usually would be only one size option available, which might or might not work based on their degree of vision loss. Many people with low vision simply read a regular printed book with a magnifier of some type, so they could have the degree of enlargement they needed. This meant relatively few books, relatively little flexibility.

With the large Bookshare library of e-books in a single primary digital format (EPUB or .epub), however, it became possible to adapt every book to the particular needs of each user and offer multiple options to the same user based on what they needed at the moment. Users with dyslexia (by far the majority of Bookshare users because there are many more people with dyslexia in the United States than with blindness) could have an audiobook, although it would be read aloud by a computer voice rather than by a human narrator. Making up for that drawback was extreme flexibility in customizing the visual appearance of the text, which could often address the unique nature of the person's reading differences.

The "killer app" for people with dyslexia turned out to be karaoke-style reading, where each word would be visually spotlighted while the voice synthesizer read it aloud ("follow the bouncing ball"). The power of bimodal reading (such as seeing and hearing the same word at the same time) had been understood in the dyslexia field for decades, but technology made it practical. Low-vision readers could adjust the text to the precise size they preferred, flip from black text on a white background to white text on a black background (very handy for people with certain eye conditions), and even switch to listening to the book if their eyes became fatigued. Blind people could get audiobooks or braille books. The braille could be printed out by a braille embosser (like an inkjet printer for sighted people but pressing the braille dots into the page instead). A huge win for braille readers was the ability to easily load hundreds of digital books into a portable braille display, which meant that an entire library of braille books could fit into a device that weighed a pound or two.

Bookshare tackled mass customization not only by making it far easier to read in the format of the user's choosing but also by increasing the number of books available to readers with disabilities by more than a factor of ten, with more than a million different accessible titles. With these advancements, readers would be more likely to have their own personal customized library of the books they wanted, available in their preferred reading format.

In the United States, every child with a disability that severely affects their education is entitled to an individualized education plan. I believe that technology advancements will make it possible over time for all children in the world to get individualized and customized educational experiences.

The attractiveness of mass customization doesn't stop with education and accessibility. In medicine, we have learned that certain cancer drugs work great for individuals with certain genetic mutations but not for others. There is also much untapped potential in improving farmer incomes and reducing climate impact from agriculture by customizing farming practices to each field. This is already being done in industrialized agriculture, where precision agriculture makes small changes in the amount of fertilizer and pesticides for specific spots in a field based on data. Tech Matters is one of a number of tech developers trying to bring the power of these approaches to all farmers in the world, not just to corn and soybean farmers in wealthier countries.

In the past, it was often seen as impractical to deliver customized services at scale to the poor and the disadvantaged. Now, the advent of affordable tech makes this kind of mass customization realistic.

### RECOMMENDATION 2: EMBRACE LARGE-SCALE ETHICAL DATA COLLECTION

Large-scale data collection is necessary to unlock opportunities that weren't possible before and to chart a better path to having an impact at scale. This applies both to listening to and enabling the people you serve as well as to using data to run your enterprises better. A key technology enabler of large-scale data collection is the cloud.

The biggest benefit of the cloud is you can use it to collect and analyze far more data than was practical in the past. Moving to the cloud also lets you leave behind the challenging days of installing software on many different types (and ages) of devices. An added benefit is that storing your applications and data in the cloud puts far more technical capabilities in the hands of anyone with access to an internet-connected device.

If you put a resource on the web, it is theoretically available to billions of people, especially with the advent of automatic translation in web browsers. For a developer of technology, this increase in reach comes with its own set of pros and cons. On the plus side, in general the cloud lowers the costs of deploying and operating your products. If you experience a sudden surge in demand, cloud providers are happy to instantly rent you far more capacity in terms of computers, storage, and bandwidth.

However, to make the broadest impact on social problems, you must not only write the software but also operate it as a cloud service because practically none of your users have the ability to set up your software in the cloud. The effort is worth it, though! Amazing leverage is possible from taking this path.

Just one example of the power of cloud-enabled large-scale data collection is Kobo Toolbox, the technology nonprofit I discussed at length in chapter 3, which enables hundreds of millions of surveys to be fielded annually by more than ten thousand organizations all over the world. Twenty years ago, that kind of scale at a budget of only $1 million a year would have been inconceivable. Kobo helps make it possible for social sector groups to ask many millions of people what they need, what they think, and what should be done.

I wrote earlier in this chapter about the benefits of shifting power over tech and data to the individuals and communities we serve and using the data to better "listen" to them and measure the impact of tech-enabled programs. Large-scale data collection does even more. It makes it easy to construct dashboards, reports, analyses, and visualizations for you, your team, and those using your technology. Beyond these, some organizations are already experimenting with conversational interfaces (enabled with AI) to allow users to simply ask questions of the database in everyday language, and get a usable answer.

When those of us in the social good field are starting a new social enterprise, we should ideally ask ourselves what we, as the leaders of the project, need to know to ensure we are being successful. And then we should make sure that we're collecting the data we need to track our progress and effectively operate our enterprise. Of course, the data being collected should be firmly based on listening to the voices of our users, which should enable us to deliver what they need from us.

We should also go one step further with the data we collect. We should use data to convince our donors and supporters to measure us in the same way we measure ourselves. That way, our team is focused on performing against one set of common objectives.

If we embrace large-scale data collection, we eventually will collect enough data to put AI to work for us as we design, deliver, and measure the success of our programs. However, large-scale data collection concentrates power, so how should we think about ensuring that we treat these data with the respect they deserve and ensure they are used to benefit the people we serve?

### A Better Deal for Data

The tech industry has figured out how incredibly valuable data are. The norm in industry is to grab all the data it can. The legal terms of service from online companies are unreadable, and yet they lay firm claim to effectively own all of the data they collect about you. The scale of these extractive data practices boggles the mind. For example, *Consumer Reports* and the journalism nonprofit the Markup published a report, based on a test group of 709 volunteers, about how more than 185,000 different companies send data to Facebook. All these companies are part of a vast ecosystem of buying, selling, and trading your data, ranging from data brokers and marketing agencies to businesses such as the Home Depot and Amazon.[7]

Such massive data collection is going on all over the tech field, and the power dynamic is such that consumers have little ability to change it. There's a popular saying in industry: "If you are not paying for it, you're not the customer; you're the product being sold."[8] In the case of platforms such as Facebook, they are selling you to advertisers. Their interest in this business model means that they act primarily in the best interests of their advertising business, not of their users. This is why divisive online content is promoted so actively by these platform companies: content that triggers strong responses leads to more engagement and time online, which lead to more profits.

It is easy to accuse large for-profit companies of bad behavior when it comes to data use. However, the communities served by the social sector have reasons to expect better. If promoting social good is our North Star, then it's reasonable to expect nonprofits and responsible businesses to commit to *not* using data to exploit people.

What does an alternative approach to data look like? Dr. LaKisha Odom from the Foundation for Food and Agriculture Research, a major US funder, is keenly interested in this topic. Odom was worried that in this time of rapid climate change, when farmer data are desperately needed to inform agriculture research, farmers are increasingly disinclined to cooperate with data collection efforts. They have begun to realize that when data are collected from them, those data are often used to their detriment.

On a Zoom call a couple of years ago, Odom challenged Steve Francis, the leader of the Terraso open-source software project at Tech Matters, and me to devise a new approach to privacy for farmers—a kind of "HIPAA for Farmers." HIPAA (Health Insurance Portability and Accountability Act of

1996) is the US medical privacy law that allows Americans to generally trust their doctors to treat them and keep their individual medical data secret while also permitting confidential access to their data for medical research. Odom wondered if such a new approach could re-create in agriculture the level of trust in data collection seen in medicine.

Steve thought we should take up Odom's challenge, so he, Dr. Katy McKinney-Bock (a data scientist also working at Tech Matters), and I started a new project to fulfill Odom's request. The working name is "A Better Deal for Data," a series of plain-language commitments around collecting personal, private, or sensitive data.[9] These commitments are intended to provide a trustworthy alternative to the surveillance capitalism norm and shift power over data to individuals and communities. Our goal is to make it easier for mission-based organizations to commit themselves and their tech vendors, if applicable, to a set of best practices. By providing organizations with the means to make straightforward public commitments to securing sensitive data and to not selling access to the data to the brokers, we hope to increase trust in tech-for-good projects built on data.

A Better Deal for Data is based on the following concepts:

- We will use your data to provide benefits to you, your community, and larger society.
- We are collecting and using your data but not claiming ownership of it.
- You can ask us to delete your data or correct it or to allow you to transfer it somewhere else.
- We will not sell your private data to third parties such as Facebook and Google.
- We will protect and steward your data and comply with data privacy laws.
- If we publish research based on your data, it will be made freely available.
- We will make legally binding promises to follow through on these commitments.

A Better Deal for Data is going through the same kind of human centered codesign processes I suggest for any tech-for-good projects (see chapters 4 and 6). Does it fit the real needs of social change organizations? How much more work is needed before it is ready to be put into practice? Iterate until it's good enough to release. We hope this will become a model for the more responsible use of data around the world.

In all events, as the leader of a tech-for-good organization, you should inform your users in some detail of what you plan to do with their data in terms they can understand *before* collecting and using it.

As I wrap up recommendation 2, the following case study involves putting more extensive data at the core of a new approach to solving a persistent social problem: homelessness.

**Community Solutions: Data Driving Change**

Social problems are notoriously difficult to solve. What happens when a celebrated leader in the fight to end homelessness decides that her approach has not been enough? Rosanne Haggerty had received a MacArthur Fellowship in recognition of her work with Common Ground, which dramatically reduced homelessness in New York City's Times Square area back when Times Square was regarded as a hopeless blight. She followed up with the 100,000 Homes project, which succeeded in its outrageously bold goal and housed more than 100,000 people across more than one hundred communities. And yet homelessness persisted in all these places.

What do you do when being highly successful in building housing for people who are homeless doesn't solve the problem? For Haggerty, you change the measure of success. Rather than continuing to focus on outputs such as housing units, her nonprofit Community Solutions shifted to measuring outcomes. The new target was "getting to zero" in a community where every person who was homeless, say, three months ago would now be housed. Once that target was achieved, you would maintain that level of impact where nobody in a community remains unhoused for long.

This was a seismic shift in approach in the field. It required the great majority of nonprofits and government agencies in a given area to work together. Organizations that confused their outputs for their mission (for example, having a mission of feeding the homeless or providing a certain number of shelter beds) would realize that getting to zero would threaten their largest programs. Rosanne described this challenge as pushing back against the "homeless industrial complex."

The change also required a completely different approach to the technology used by these organizations. A big barrier to overcome was that most homelessness funding in the United States comes from the federal government, which mandates the use of the Homeless Management Information System (HMIS) in any given community using federal funding. The HMIS approach was designed back at the turn of the twenty-first century and is more focused on measuring and tracking outputs than on eliminating homelessness. Apparently, it's quite cumbersome to use and not particularly helpful, as any piece of software dating back a couple of decades would likely to be. Haggerty shared with me that as she was initially coming up with the Built for Zero idea, many leaders in the field felt that the HMIS mandate was one of the biggest barriers to ending homelessness in the country. As a tech guy, I was shocked. I had never before heard that the legacy tech in a field was blocking progress toward what should be everybody's systems change goal: eliminating homelessness!

The short-hand description of what Haggerty and her team think is needed instead is a "by-name database." To know if everyone who was homeless two months ago is now housed, you will need to track every person who has been experiencing homelessness in a community across dozens of organizations—a very difficult challenge. However, this is now technically possible and creates opportunities for data to offer a new approach to homelessness.

Of course, tracking individuals and sharing their data confidentially across many organizations raises privacy concerns, although I have found no instances so far where this information has been misused (or shared with law enforcement). It is to be hoped that the coalition of groups working toward the social goal of ending homelessness in their communities will continue to honor the social bargain of using the data to assist the people who need help, not for other purposes.

So far the federal mandate has meant that communities working toward zero need to have two tech systems: one legacy system meeting the HMIS requirements and a by-name database for measuring progress toward zero. Someday, maybe, it will be possible to build a single modern system to collect the data that meet both sets of requirements, but such a project will probably take an act of Congress. Meanwhile, efforts are focusing on getting these systems to work better together through improved application programming interfaces and data standards.

Haggerty and Community Solutions got quite a boost in 2021 when they won the MacArthur Foundation's 100&Change competition, which granted $100 million to their Built for Zero concept.[10] The goal of Built for Zero's MacArthur proposal was to get seventy-five communities to "functional zero," where they will have effectively reached this new target of eliminating persistent homelessness. This level of funding should be enough to try out this innovation at scale.

So far Built for Zero is making good progress, especially in smaller and medium-sized metro areas. Its ambitious (and measurable) goal is to drive changes in homelessness programs in favor of those that "get to zero." For example, the "housing first" approach, where homeless people are not required to meet certain requirements (say, to stop using alcohol or other drugs) to be housed, is becoming more popular in these communities. The question is still open about whether this approach can succeed in a large city with its more challenging politics and affordability problems.

Of course, a complex social problem such as homelessness cannot be solved solely by a new database and better data quality! Community Solutions' new approach should be a source of better information about getting to zero and identifying the kinds of resources and programs that are the most successful in reaching that target. Brian Trelstad of Harvard Business School did a case study on Haggerty and Community Solutions, and in an interview he mused about whether her work, as well as better and more sophisticated data, could succeed in changing the way the federal government makes policy and funding decisions about homelessness.[11]

The shift that Rosanne Haggerty initiated is a great example of one of the best ideas in tech for good that I described in chapter 3: killing the dinosaurs. This tactic is often necessary because the existing approaches to ending homelessness have clearly fallen short. Not only is Haggerty trying to change the entire field, but she is also moving past her initial work around building housing units. If she and her Built for Zero coalition succeed, there will be less need for status quo approaches to homelessness, which will have the side effect of reducing funding for and employment in these programs. However, it is clear that the right goal for society is not to perpetuate homelessness programs but to make major progress in eliminating homelessness itself.

## RECOMMENDATION 3: USE AI FOR GOOD

Large-scale data collection has its biggest impact when it is used to train effective AI tools. Even as I have cautioned throughout this book against bad applications of tech and AI, I strongly believe that AI tools will be increasingly indispensable in the social sector in the coming years. AI is critical to the future of social change. But that future needs smart AI applications, not dumb ones!

Building smart AI applications is not easy. I believe most AI-for-good efforts currently fail 90–95 percent of the time if success is measured as delivering an exciting return on the funds invested in building the tech. This is a common consequence of the Gartner Hype Cycle, which overstates the benefits of a new technology, as in the case of advanced AI. When nonprofits rush to apply a hot technology to a real problem, the tech is frequently not capable of delivering significant benefits greater than the status quo. Having many, many nonprofits each hiring expensive data science resources to tackle a common problem means that the funding dedicated to AI applications is spread over a large number of projects, and there is a lack of critical mass. Far better to collaborate or wait for better AI products!

In my earlier chapter on why AI projects in the social change world fail, I noted the following:

- Computers are still dumb as bricks, even if they appear to be smart.
- AI makes mistakes. It is never perfect.
- The data on which AI is based are biased. Full stop. The world is biased; AI is trained on data from the world; therefore, AI is biased.

So what's an optimistic social leader to do? I advise you to keep the limitations of AI in mind, design around them, and forge ahead with plans to make the power of data and algorithms (a.k.a. AI) effectively play a role in your efforts if those plans seem likely to be a cost-effective way to deliver better impact. As the data scientist Cathy O'Neil points out in her book *Weapons of Math Destruction* (2016), mentioned previously in this book, you need to make sure your algorithms are optimized for what your community defines as success, not for what the existing power structure prioritizes.

I also advise that you never put the robot in charge (remember that computers are still dumb as bricks). The AI system doesn't understand the answers it hands out, recognize when it is wrong, or know what the impact of being wrong is or will be. I've observed an unfounded deference to advanced technology systems by sensible people who turn over their decisions to a machine. Decision-making is not that easy. Today, we still need humans to catch and fix the errors coming out of the machine.

Of course, the sales pitches from the tech field often bound on irrational exuberance, without much openness to challenge. Technologists like to wave terms such as *artificial intelligence* at nontechnologists. To reference O'Neil again, implying that a technology or advanced system is too hard for mere mortals to understand is often a way to shut down conversations about the effectiveness of the technology or system. If the pros and cons of an AI system or how it works are too hard for nontechnologists to understand, it's probably a bad idea to implement such a system!

There is only so much AI can do for the social sector. If you are a social leader who is an expert in your field but not in AI, remember that *you* are the expert in the field. An AI solution will never know as much about your field as you do. If someone is pitching you on the glories of their AI tool, ask a lot of questions. If you can't see quickly how a tech tool can be beneficial to your work, it probably won't be. A tool that will save you valuable time and make your team more effective is well worth a serious look, and the best place to find such tools is from peers who have successfully applied them.

Here are some other guidelines that will help you find a solid path for applying AI capabilities to your social mission and avoid some of the many pitfalls.

First, today's AI miracle is tomorrow's workhorse. My recommendation? Wait for tomorrow. Wait until the latest hot fad has been tried out by the tech people at the leading edge. Let the venture capitalists and for-profit companies burn billions of dollars and both succeed and fail at figuring out what AI applications work well. Wait for the costs to come down. Unless you've decided to make a major investment (for instance, $1 million or more a year), maybe the best way to start is to support a joint AI-enabled open-source project in your field. Or wait for a product that has been designed to take advantage of AI's capabilities and mitigate its shortcomings. Finally, keep an eye on well-funded tech-for-good nonprofits to see how this is done.

Two exciting examples of leading-edge AI applications are Digital Green and Khanmigo. In both cases, developers used generative AI and put in an effort to make it work well in a narrow field (rather than using an open-ended chatbot such as ChatGPT). Digital Green is a tech social enterprise that spun out of Microsoft Research and focuses on helping smallholder farmers be more successful. The organization has developed a huge amount of video content demonstrating best practices, all featuring real situations and people speaking the local languages of farmers. A team of roughly thirty people trained an AI model with content in a handful of languages, including Hindi for farmers in parts of India and Amharic for farmers in Ethiopia. The team worked for two years and after a great deal of experimentation ended up with a worthwhile tool. Even so, the tool was not initially made available directly to farmers but was instead provided to agriculture extension agents, who were expected to catch the remaining AI tool's mistakes before putting it in the hands of farmers.

Khanmigo is from Khan Academy, a well-known tech-for-good educational nonprofit with a particularly strong tech team. Khanmigo is a chatbot tutor paired with Khan Academy's extensive educational content library, designed to guide learners to learn rather than giving them the answers to their questions. It uses the current OpenAI technology, which is expensive. Khan can't make Khanmigo available for free (unlike its educational videos), but it has worked to get the cost down. I am excited to see how successful Khanmigo will be in making an impact on students over the coming years.

Notwithstanding these two examples, it is very difficult for nonprofit organizations with a deep responsibility to disadvantaged communities to use valuable donor funding to run experiments on a technology that only *might* be helpful to the disadvantaged. Remember my blockchain example in chapter 2. How much social sector funding was wasted along the way to nowhere?

Second, rather than putting the robot in charge, keep the humans in the driver's seat. AI tools can be great for making smart humans smarter and more productive. Instead of laying off a bunch of people and settling for whatever stuff the robot spits out, wouldn't it be better if your people became 25 percent more productive without introducing more errors? I have already discussed how thoughtlessly outsourcing responsibility to the robot (or equivalent) is a bad idea. But we must also remind ourselves that there are so many ways AI can help advance social impact with mindful design that keeps humans in the loop.

Third, pick applications that are meaningful, where there will be strong benefits but also where errors are not costly. Recall the positive AI examples from the crisis response helpline field in chapter 5. Instead of firing all your workers and replacing them with a chatbot (as in the case of NEDA and its chatbot, Tessa)—which will do the opposite of a good job—you should think about ways to strengthen the people in the system you are working to improve.

There are plenty of other useful applications for AI. Of course, it may be that the cost of implementing one of them may exceed the benefits over time. This is where ideas such as total cost of ownership (TCO) come in. If you spend a year and $200,000 of donor funding on an AI application project, is it a better use of the effort and funding than something else? Will it save you $500,000 or more in your team's time over the next five years? If not, you might not want to invest in the effort because tech development is never a sure thing!

## CONCLUSION

As you chart the future of your tech-for-good efforts, please keep in mind the stories and suggestions I have shared in this book. By sensibly embracing the unstoppable progress of technology, you should have the opportunities to make the most positive social change.

In particular, by using technology to change the unjust power relationships dominating industry's use of data, we can help ensure that individuals and communities can flourish in the face of exploitation. Large-scale data collection, if done with ethical values and local empowerment in mind, can be indispensable for hearing the voices of society's most disadvantaged and designing programs that truly benefit them. The immense potential of AI should be harnessed to serve all of humanity.

# AFTERWORD: GO FORTH AND USE TECH FOR GOOD!

In this book, I have explained why the future of nonprofit efforts over the next decade will be built on the better use of technology. I have advised that this shift will require equipping the typical nonprofit with modern tech, creating new tech social enterprises to provide cloud solutions for entire fields, and unleashing dinosaur killers to drive innovation into fields resistant to change. I have tried to make the intangible and often impenetrable nature of technology real for you, readers who care about doing more for people.

I hope I have convinced you that tech matters. I hope you have been inspired by the stories I've shared of social leaders who have been successful in making an impact.

It's time for you to act.

- Technologists can lend their knowledge and skill in tech to create positive change of their own.
- Nonprofit leaders can use technology for far greater impact in their chosen fields.
- Social good tech entrepreneurs can be reassured that they are not alone and emboldened to follow their own path, assisted by this book's suggested roadmap and the stories of their peers.
- For-profit tech company leaders can provide social good entrepreneurs with discounted access to their products, so their creations can benefit all of humanity, not just the wealthiest. And when these for-profit leaders get rich (!), they should seek out the best-performing nonprofits to deploy the philanthropic assets of their companies and themselves.

- Philanthropists can find their work illuminated by the light of better data and amplified by the power technology provides to their grantees, and they can feel confident about supporting nonprofits to use technology for the long term.
- Policymakers can enact more enlightened regulation of tech to encourage its beneficial uses and to mitigate its negative effects, and they must recognize how staid government programs can be supercharged by the right technology adoption.
- Students can pursue careers in which technology is focused primarily on helping human beings and the planet.
- All readers can join me and the growing numbers of social tech-for-good leaders in the shared goal of harnessing the power of tech for people, not profit.

Now, go forth and make change!

# Acknowledgments

This book, like a social enterprise, is the culmination of the work of so many people who had a hand in shaping a successful outcome. I want to thank in particular Susan Thomas, my developmental editor, who helped channel my desperate longing to spread the word about tech for good into a sufficiently strong proposal that prompted my agent, Joe Spieler, to ask, "What happened?!" Joe's willingness to stick with me through fifteen years of fits and starts has demonstrated great patience.

I am honored that Vilas Dhar, one of the leading voices on using technology and AI for social impact, graciously agreed to write the foreword for this book. Vilas leads the Patrick J. McGovern Foundation, which in just a few short years has established itself as the leading tech-for-good funder in the field, willing to take bold risks on exciting ideas (including some of ours at Tech Matters).

In support of the book process, I had the benefit of three Tech Matters fellows, Dr. Katy McKinney-Bock, Gabriele Carotti-Sha, and Erin Brennan, who together found tech-for-good enterprises I needed to know about, located authoritative references for facts I thought I was making up, and otherwise helped me refine the stories and themes you see in this book. Gabriele also helped me launch the *Tech Matters Podcast* and continues to produce it years after the ending of his theoretically one-year fellowship. I also thank the Rockefeller Foundation and the team at the Bellagio Center for supporting me with a Bellagio residency, where I was able to complete a first draft of this manuscript. My fellow residents in Bellagio, all highly accomplished leaders and authors, were a font of inspiration, feedback, and practical advice on publishing a book.

I thank all of the great tech-for-good leaders who invested their time in speaking with me over many years: their innovations are at the heart of all of the lessons I hope to teach with this work.

My two social enterprises, Benetech and Tech Matters, have been successful because of contributions by so many outstanding people who shared the dream that technology can be channeled into delivering social good. Unlike in the writing of a book, the great majority of the work to create a successful social enterprise is not done by the person who happened to be there at the beginning of the journey, even though our field's approach to awards focuses on naming founders rather than teams. Benetech and Tech Matters have succeeded because of the people who powered that impact through their day-to-day work. And because of the outstanding leaders who volunteered to serve on our boards of directors, supporting and guiding me and our teams. Of course, so much of what we accomplished was made possible with the help of the risk-taking donors who placed their faith in our teams. I regret that I only have space here to acknowledge a few of the many supporters who funded our work over multiple decades.

Among those who deserve credit for helping me build two successful nonprofits are the boomerangers, people who chose to rejoin my teams after being teammates in an earlier stage in my career. Chief among them is Joan Mellea, the cofounder of Tech Matters and my right-hand person at Benetech for many years. Joan helped me make space for this book while we were scaling up Tech Matters and convinced me that the most compelling book to write would spotlight the innovations of dozens of other outstanding tech-for-good organizations.

Jane Poole and Teresa Throckmorton played the indispensable roles of leading people and finance at Benetech, reprising roles they had in my first successful startup, Calera (where a board veto ironically launched my tech-for-good career). Bill Schwegler was poached from Calera to run product and engineering for the Arkenstone products, after having co-led the development of the reading machine prototype with Dave Ross, my Benetech cofounder (and cofounder with me of RAF Technology, another successful

AI startup). I worked for Dave on the Percheron Project, the first private-enterprise rocket venture to get a rocket to the launch pad, where of course it promptly blew up (a story beyond the scope of this present work)! Steve Francis, who as a Stanford undergrad worked for me on the rocket project, had a successful entrepreneurial career in Silicon Valley and joined me at Tech Matters to run our Terraso climate change project. Celine Takatsuno has joined us as our latest fellow to work on A Better Deal for Data, more than a decade after helping launch a social enterprise at Benetech. And, finally, there is Aaron Firestone, who after more than a decade of working with Joan Mellea and me at Benetech joined Tech Matters to help replace the irreplaceable Joan when she retired.

Of course, I haven't (yet) had the chance to work a second time with many other great teammates who made Tech Matters and Benetech successful. Nick Hurlburt and Dee Luo are the dynamic duo who led the Aselo crisis response platform at Tech Matters to great success. Betsy Beaumon, Benetech's longtime president and my successor as CEO, oversaw the growth and success of Bookshare after we secured the major federal funding infusion that saved that project. A long-lived social enterprise, Bookshare has had many talented leaders, including Alison Lingane, Jesse Fahnestock, Janice Carter, Lisa Friendly, and Brad Turner.

Leading an outstanding, world-changing organization like Benetech for thirty years means that I benefited from interactions with many more people who played critical roles in its success. I particularly want to acknowledge Christy Chin, who was my first program officer at the Skoll Foundation and went on to become an incredible long-serving Benetech board chair. Clare Hamm Geoffray was Benetech's very first employee (back when we were still named Arkenstone) and laid the foundations for Benetech's culture. Roberta Brosnahan was Arkenstone's chief operating officer and did so much to ensure its success, including driving the user survey that surfaced how many students with dyslexia were using our reading machine for the blind. Patrick Ball, one of the world's top human rights data scientists, played the crucial role in Benetech's work in human rights and continues that work to this day as the cofounder of the Human Rights Data Analysis Group, which spun out of Benetech years ago.

Bookshare's success has depended in particular on leadership from key players outside our organization. George Kerscher, the brilliant technologist and key person behind the modern e-book format EPUB, created the technology that made Bookshare possible and founded the very first accessible e-book nonprofit in the 1980s. Glinda Hill and Larry Wexler of the Office of Special Education Programs at the US Department of Education took huge risks by betting on the Bookshare team and defended us through multiple stressful periods (a DC staple). Allan Adler, the general counsel of the Association of American Publishers, made the choice to work closely with me to help make Bookshare possible, while also protecting the interests of publishers and people with disabilities. And Donna McNear, the foresightful teacher of the people with visual impairments in rural Minnesota, encouraged me to go to Washington, DC, to visit Glinda Hill at the Department of Education despite my reservations.

No acknowledgments would be complete without a mention of my mentors and my family, who helped me be a better leader and a better person. Chief among my mentors was the late great Gerry Davis, who was behind the founding of Benetech (he did the pro bono work to found it as a charity) and shaped the legal strategy that made Bookshare possible. When you find a problem-solving attorney like Gerry, instead of the more typical problem-creating ones, you should never let go of them! Jim Kleckner is still my alpha geek, the engineer's engineer who explains new technology to me as it comes out and helps me figure out how to use it for good. Ben Rosen, the original technology pundit and venture capitalist, constantly took risks to support my work, starting with investing in my first for-profit AI company, Calera, and continuing to mentor me well into his nineties!

Jeff Skoll, eBay's original president and longtime supporter of Benetech and Tech Matters, showed me how an engineer can go on to help accomplish great social good work, from founding the Skoll Foundation to creating and leading Participant Media, which brought so many social change stories to Hollywood. Professor Bert Hesselink was my key mentor at Caltech, and it was in his pattern recognition lecture that I first came up with the idea of using technology to help blind people read books on their own. My original mentors were the clerics of St. Viator High School, who showed me how a love of service could be combined with a love of math, science, and software development. This approach to life was exemplified by Father Arnie Perham, whose fateful choice to put the Caltech Beaver cheer—the finale to my acknowledgments—on the wall of his math classroom sent me to Pasadena!

Last in this list but first in my heart are my wife, Virginia, and our three children—Jimmy, Andy, and Kate. Their support for my nonprofit career, an undoubtably less lucrative path for our family, is what kept me going in tech for good for more than thirty-five years. Now with granddaughter Juniper on the scene, I have a renewed commitment to try to build a better world with technology for the benefit of Juniper's generation.

Cosine, tangent, hyperbolic sine,
Three point one four one five nine
E to the x, DY DX,
Slide rule, slipstick, Tech, Tech, Tech!

# Notes

## PREFACE

1. All dollar amounts given throughout the book are US dollars.

## INTRODUCTION

1. Jim Fruchterman, host, *Tech Matters Podcast*, produced by Gabriele Sha and sound edited by Phil Kadet, accessed June 19, 2024, https://techmatters.org/tag/tech-matters-podcasts/.

## CHAPTER 1

1. Skoll Foundation, "Skoll Awardees," accessed May 28, 2024, https://skoll.org/community/awardees/. This percentage was determined by creating a spreadsheet of all Skoll awardees from 2018 to 2023, manually assessing which organizations develop their own technology, and then calculating that proportion relative to all awardees.
2. Kobo, "KoboToolbox 2023 Year in Review: Global Impact Highlights," KoboToolbox, March 20, 2024, https://www.kobotoolbox.org/blog/kobotoolbox-2023-year-in-review-global-impact-highlights; "About the Kobo Organization," KoboToolbox, accessed July 14, 2023, https://www.kobotoolbox.org/about-us/the-organization/.

## CHAPTER 2

1. Olivia Sidoti et al., "Mobile Fact Sheet," Pew Research Center, January 31, 2024, https://www.pewresearch.org/internet/fact-sheet/mobile/.
2. Andrew Chen, "New Data Shows Losing 80% of Mobile Users Is Normal, and Why the Best Apps Do Better," *Andrew Chen Archives* (blog), accessed June 12, 2024, https://andrewchen.com/new-data-shows-why-losing-80-of-your-mobile-users-is-normal-and-that-the-best-apps-do-much-better/.
3. Salem Solomon, "South African–Created Mobile Alert Puts COVID Info into Hands of Millions," Voice of America, April 4, 2020, https://www.voanews.com/a/africa_south-african-created-mobile-alert-puts-covid-info-hands-millions/6186959.html.

4. Tanya O'Carroll, Danna Ingleton, and Jun Matsushita, "Panic Button: Why We Are Retiring the App," *The Engine Room* (blog), September 1, 2017, https://www.theengineroom.org/panic-button-retiring-the-app/; Tanya O'Carroll, Danna Ingleton, and Jun Matsushita, "Panic Button: Lessons for the Tech for Good Sector," *The Engine Room* (blog), September 1, 2017, https://www.theengineroom.org/panic-button-lessons-learned/.

5. Jim Fruchterman, host, *Tech Matters Podcast*, season 1, episode 7, "Emily Jacobi: Creating Tools to Defend Human Rights," produced by Gabriele Sha, January 10, 2022, https://techmatters.org/podcast-episode-7-emiliy-jacobi-creating-tools-to-defend-human-rights/.

6. Jim Fruchterman, host, *Tech Matters Podcast*, season 1, episode 1, "Gamification That Works: Our Chat with Michael Sani," produced by Gabriele Sha, February 18, 2021, https://techmatters.org/gamification-that-works-our-chat-with-michael-sani/.

7. Luke Jordan, *Don't Build It: A Guide for Practitioners in Civic Tech / Tech for Development* (Capetown, South Africa: Grassroot; Cambridge, MA: MIT Governance Lab, 2021), https://mitgovlab.org/wp-content/uploads/2021/04/Grassroot-MIT_Dont-Build-It_English.pdf.

8. Cherry Yang, "What Happened to AbleData.Com?," *Mobility with Love* (blog), June 29, 2021, https://www.mobilitywithlove.com/what-happened-to-abledata/.

9. Amelia Hoover Green, "Multiple Systems Estimation: Does It Really Work?," *Human Rights Data Analysis Group (HRDAG)* (blog), March 26, 2013, http://hrdag.org/2013/03/26/mse-does-it-really-work/.

10. Wikimedia Foundation, *Annual Report 2022–2023* (San Francisco: Wikimedia Foundation, 2023), https://wikimediafoundation.org/news/reports/2022-2023-annual-report/.

11. Jim Fruchterman, host, *Tech Matters Podcast*, season 2, episode 7, "Mapping the Global Supply Chain, with Natalie Grillon of Open Supply Hub," produced by Gabriele Sha, April 19, 2024, https://techmatters.org/mapping-global-supply-chain-natalie-grillon-open-supply-hub/.

12. "Gartner Hype Cycle Reviews Digital Technology and Trends," Gartner, accessed June 12, 2024, https://www.gartner.com/en/marketing/research/hype-cycle.

13. Jake Porway, "Breaking the Cycle," keynote presented at the Good Tech Fest 2024, May 28, 2024, https://www.goodtechfest.com/recordings/v/gs9y9ak6zdgfefeeg7khf9ph7zhj2r.

14. Russ Juskalian, "Inside the Jordan Refugee Camp That Runs on Blockchain," *MIT Technology Review*, April 12, 2018, https://www.technologyreview.com/2018/04/12/143410/inside-the-jordan-refugee-camp-that-runs-on-blockchain/.

15. Cryptopedia Staff, "What Was the DAO?," Cryptopedia by Gemini, October 5, 2023, https://www.gemini.com/cryptopedia/the-dao-hack-makerdao.

16. Jemimah Jones, "Largest Cryptocurrency Hacks in History: How They Happened," *CoinCentral*, October 24, 2023, https://coincentral.com/largest-cryptocurrency-hacks-in-history-how-they-happened/.

## CHAPTER 3

1. Jim Fruchterman, host, *Tech Matters Podcast*, season 1, episode 2, "Ana Pantelic: Designing Tech for Low-Income Families in the Global South," produced by Gabriele Sha, May 6, 2021, https://techmatters.org/podcast-episode-2/.

2. "Statistics about Sexual Violence," National Sexual Violence Resource Center, Info & Stats for Journalists, 2015, https://www.nsvrc.org/sites/default/files/publications_nsvrc_fact sheet_media-packet_statistics-about-sexual-violence_0.pdf, citing Christopher P. Krebs et al., *The Campus Sexual Assault (CSA) Study: Final Report* (Rockville, MD: National Criminal Justice Reference Service, 2007), https://www.ojp.gov/pdffiles1/nij/grants/221153 .pdf.

3. Jim Hopper, "Repeat Rape by College Men," accessed May 28, 2024, https://jimhopper .com/topics/sexual-assault-and-the-brain/repeat-rape-by-college-men/.

4. National Institute of Justice, "Most Victims Know Their Attacker," September 30, 2008, https://nij.ojp.gov/topics/articles/most-victims-know-their-attacker.

5. Clayton Christensen, *The Innovator's Dilemma: When New Technologies Cause Great Firms to Fail* (Cambridge, MA: Harvard Business Review Press, 1997).

6. Kevin Starr, "Don't Feed the Zombies," *Stanford Social Innovation Review*, March 28, 2023, https://ssir.org/articles/entry/dont_feed_the_zombies.

7. Kobo, "KoboToolbox 2023 Year in Review."

## CHAPTER 4

1. Dana O'Donovan and Noah Rimland Flower, "The Strategic Plan Is Dead. Long Live Strategy," *Stanford Social Innovation Review*, January 10, 2013, https://ssir.org/articles/entry /the_strategic_plan_is_dead._long_live_strategy.

2. Leandra Fernandez, "Empathy and Social Justice: The Power of Proximity in Improvement Science," Carnegie Foundation blog, April 21, 2016, https://www.carnegiefoundation.org /blog/empathy-and-social-justice-the-power-of-proximity-in-improvement-science/.

3. "Public Interest Technology University Network," accessed September 16, 2024, https:// pitcases.org/; Catherine D'Ignazio and Lauren F. Klein, *Data Feminism* (Cambridge, MA: MIT Press, 2020).

4. Jim Fruchterman, host, *Tech Matters Podcast*, season 2, episode 6, "Bridging the Education Gap: TalkingPoints' Heejae Lim on Tech-Driven Family Engagement," produced by Gabriele Sha, April 6, 2024, https://techmatters.org/education-gap-talking-points-heejae -lim/.

5. Jim Fruchterman, host, *Tech Matters Podcast*, season 1, episode 5, "Sanjay Purohit: Designing for (Massive) Scale," produced by Gabriele Sha, October 11, 2021, https://techmatters.org /podcast-episode-5-sanjay-purohit-designing-for-massive-scale/.

## CHAPTER 5

1. Cathy O'Neil, *Weapons of Math Destruction: How Big Data Increases Inequality and Threatens Democracy* (New York: Crown, 2016).
2. Cathy O'Neil, interview by the author, Bellagio, Italy, July 26, 2023.
3. Emily M. Bender et al., "On the Dangers of Stochastic Parrots: Can Language Models Be Too Big? 🦜," in *Proceedings of the 2021 ACM Conference on Fairness, Accountability, and Transparency* (New York: ACM, 2021), 616–617.
4. Mike Solomon, "GPT3 Is Just Spicy Autocomplete," *The Cleverest* (blog), February 2, 2023, https://thecleverest.com/gpt3-is-just-spicy-autocomplete/.
5. Jeffrey Dastin, "Insight—Amazon Scraps Secret AI Recruiting Tool That Showed Bias against Women," Reuters, October 11, 2018, https://www.reuters.com/article/idUSKCN1MK0AG/.
6. Leonardo Nicoletti and Dina Bass, "Humans Are Biased. Generative AI Is Even Worse," *Bloomberg Technology + Equality*, June 9, 2023, https://www.bloomberg.com/graphics/2023-generative-ai-bias/.
7. Julia Angwin et al., "Machine Bias," ProPublica, May 23, 2016, https://www.propublica.org/article/machine-bias-risk-assessments-in-criminal-sentencing.
8. Joy Buolamwini, "When the Robot Doesn't See Dark Skin," *New York Times*, June 21, 2018, https://www.nytimes.com/2018/06/21/opinion/facial-analysis-technology-bias.html; Joy Buolamwini and Timnit Gebru, "Gender Shades: Intersectional Accuracy Disparities in Commercial Gender Classification," in *Proceedings of the 1st Conference on Fairness, Accountability, and Transparency* (New York: ACM, 2018), 77–91.
9. Arvind Narayanan and Sayash Kapoor, *AI Snake Oil: What Artificial Intelligence Can Do, What It Can't, and How to Tell the Difference* (Princeton, NJ: Princeton University Press, 2024).
10. O'Neil, *Weapons of Math Destruction*, 143.
11. Sara Merken, "New York Lawyers Sanctioned for Using Fake ChatGPT Cases in Legal Brief," Reuters, June 26, 2023, https://www.reuters.com/legal/new-york-lawyers-sanctioned-using-fake-chatgpt-cases-legal-brief-2023-06-22/.
12. Jim Fruchterman, "Generative AI Is All about the Money," *Stanford Social Innovation Review*, January 25, 2024, https://ssir.org/articles/entry/generative-ai-openai-all-about-the-money.
13. Beth Kanter and Allison H. Fine, *The Smart Nonprofit: Staying Human-Centered in an Automated World* (New York: Wiley, 2022).
14. "MapBiomas Brasil," accessed June 13, 2024, https://brasil.mapbiomas.org/en/o-projeto/.
15. C. M. Souza et al., "Reconstructing Three Decades of Land Use and Land Cover Changes in Brazilian Biomes with Landsat Archive and Earth Engine," *Remote Sensing* 12, no. 17 (2020): art. 17, https://doi.org/10.3390/rs12172735.
16. Abbie Harper, "A Union Busting Chatbot? Eating Disorders Nonprofit Puts the 'AI' in Retaliation," *Labor Notes*, May 4, 2023, https://www.labornotes.org/blogs/2023/05/union-busting-chatbot-eating-disorders-nonprofit-puts-ai-retaliation.

17. Kate Wells, "An Eating Disorders Chatbot Offered Dieting Advice, Raising Fears about AI in Health," NPR, June 9, 2023, https://www.npr.org/sections/health-shots/2023/06/08/1180838096/an-eating-disorders-chatbot-offered-dieting-advice-raising-fears-about-ai-in-hea; Catherine Thorbecke, "National Eating Disorders Association Takes Its AI Chatbot Offline after Complaints of 'Harmful' Advice," CNN, June 1, 2023, https://www.cnn.com/2023/06/01/tech/eating-disorder-chatbot/index.html.

18. Andrew Myers, "AI-Detectors Biased against Non-native English Writers," *Stanford University Human-Centered Artificial Intelligence (HAI)* (blog), May 15, 2023, https://hai.stanford.edu/news/ai-detectors-biased-against-non-native-english-writers; Weixin Liang et al., "GPT Detectors Are Biased against Non-native English Writers," *Patterns* 4, no. 7 (July 2023): art. 100779, https://doi.org/10.1016/j.patter.2023.100779.

19. For an overview of the AI Act, see "EU AI Act: First Regulation on Artificial Intelligence," *European Parliament* (blog), August 6, 2023, https://www.europarl.europa.eu/topics/en/article/20230601STO93804/eu-ai-act-first-regulation-on-artificial-intelligence.

20. Nicoletti and Bass, "Humans Are Biased. Generative AI Is Even Worse."

21. Jim Fruchterman and Joan Mellea, "Expanding Employment Success for People with Disabilities," Benetech, November 2018, https://benetech.org/wp-content/uploads/2018/11/Tech-and-Disability-Employment-Report-November-2018.pdf.

22. Isaiah Poritz, "OpenAI Faces Existential Threat in *New York Times* Copyright Suit," *Bloomberg Law*, December 29, 2023, https://news.bloomberglaw.com/ip-law/openai-faces-existential-threat-in-new-york-times-copyright-suit.

23. David Thiel, "Investigation Finds AI Image Generation Models Trained on Child Abuse," *All Cyber News* (blog), Stanford Cyber Policy Center, December 20, 2023, https://cyber.fsi.stanford.edu/news/investigation-finds-ai-image-generation-models-trained-child-abuse.

24. Alexandra S. Levine, "Suicide Hotline Shares Data with For-profit Spinoff, Raising Ethical Questions," *Politico*, January 28, 2022, https://www.politico.com/news/2022/01/28/suicide-hotline-silicon-valley-privacy-debates-00002617.

25. Jim Fruchterman, "The Future of Tech for Child Helplines," Tech Matters, October 19, 2022, https://techmatters.org/the-future-of-tech-for-child-helplines/.

26. Trevor News, "The Trevor Project Expands Its AI Innovation with New 'Crisis Contact Simulator' Persona to Scale Counselor Training," *The Trevor Project* (blog), December 7, 2021, https://web.archive.org/web/20211222231230/https://www.thetrevorproject.org/blog/the-trevor-project-expands-its-ai-innovation-with-new-crisis-contact-simulator-persona-to-scale-counselor-training/.

## CHAPTER 6

1. Jim Fruchterman, "For Love or Lucre," *Stanford Social Innovation Review*, Spring 2011, https://ssir.org/articles/entry/for_love_or_lucre.

2. Ann Mei Chang, *Lean Impact: How to Innovate for Radically Greater Social Good* (New York: Wiley, 2018), 39–52.

3. Meredith Broussard, *Artificial Unintelligence: How Computers Misunderstand the World* (Cambridge, MA: MIT Press, 2019).

4. Kate Orkin, "The Evidence Behind Putting Money Directly in the Pockets of the Poor," *Oxford News Blog*, University of Oxford, May 12, 2020, https://www.ox.ac.uk/news/science-blog/evidence-behind-putting-money-directly-pockets-poor.

5. Jim Fruchterman, "Landmine Detector Project Lessons Learned," *Benetech* (blog), December 2, 2007, https://benetech.org/blog/landmine-detector-project-lessons-learned/.

6. "Martus Sunsets Software for Human Rights Data Collection," *Benetech* (blog), May 15, 2018, https://benetech.org/blog/martus-sunsets-human-rights-data-collection/.

7. "Frontline Is Closing," FrontlineSMS, June 28, 2021, https://www.frontlinesms.com/blog/2021/6/28/frontlinesms-is-closing.

## CHAPTER 7

1. Alexandra Lilienfeld, "Nexleaf Analytics Receives $12M from MacKenzie Scott," Nexleaf Analytics, March 23, 2022, https://nexleaf.org/nexleaf-analytics-receives-12m-from-mackenzie-scott/.

2. "TechSoup's Direct Public Offering," September 18, 2019, http://www.techsoup.org/direct-public-offering.

3. "About the Giving Pledge," Giving Pledge, accessed June 13, 2024, https://www.givingpledge.org/about.

4. Tara Dawson McGuinness and Hana Schank, *Power to the Public: The Promise of Public Interest Technology* (Princeton, NJ: Princeton University Press, 2021).

5. "FAQs & Support," American Youth Soccer Organization (AYSO), accessed June 2, 2024, https://ayso.org/faqs-support/.

6. Quoted in Philip Ngari, "Partners in Health to Open Source Their Full CHT-Based Community Health App—YendaNafe," *Medic* (blog), October 18, 2022, https://medic.org/stories/partners-in-health-to-open-source-their-full-cht-based-community-health-app-yendanafe/.

7. "Commoner Letter #2—Cory Doctorow," posted by melissa on the Creative Commons website, October 29, 2007, https://creativecommons.org/2007/10/29/commoner-letter-2-cory-doctorow/.

## CHAPTER 8

1. Russ Roberts, host, "Mauricio Miller on Poverty, Social Work, and the Alternative," *EconTalk*, podcast, May 6, 2019, https://www.econtalk.org/mauricio-miller-on-poverty-social-work-and-the-alternative/.

2. Shoshana Zuboff, *The Age of Surveillance Capitalism* (New York: PublicAffairs, 2019).

3. Nithya Ramanathan et al., "Decolonize Data," *Stanford Social Innovation Review*, Spring 2022, https://ssir.org/articles/entry/decolonize_data.

4. Jono McKay, "Chatting for Change," keynote presented at the Good Tech Fest 2024, May 8, 2024, https://www.goodtechfest.com/recordings/v/4lk4fw39d9yyj45l34tkhwcs9t6wfr.

5. Ellie Yu, "Kids Help Phone Crisis Texting Service Help Young People Avoid Emergency Department Visits," *Canada Health Infoway* (blog), November 20, 2019, https://www.infoway-inforoute.ca/en/news-events/blog/innovation/kids-help-phone-crisis-texting-service-help-young-people-avoid-emergency-department-visits.

6. Jim Fruchterman, host, *Tech Matters Podcast*, season 2, episode 1, "Overcoming the Market (Failure), with Owen Barder of PxD—Part 1," produced by Gabriele Sha, January 26, 2024, https://techmatters.org/podcast-launching-season2-owen-barder-part1/.

7. Jon Keegan, "Each Facebook User Is Monitored by Thousands of Companies," *Consumer Reports*, January 17, 2024, https://www.consumerreports.org/electronics/privacy/each-facebook-user-is-monitored-by-thousands-of-companies-a5824207467/.

8. Will Oremus, "Are You Really the Product?," *Slate*, April 27, 2018, https://slate.com/technology/2018/04/are-you-really-facebooks-product-the-history-of-a-dangerous-idea.html.

9. "A Better Deal for Data: Join Us in Developing Guidelines for Collecting, Storing and Using Data Better," A Better Deal for Data, accessed June 13, 2024, https://bd4d.org/.

10. "Community Solutions Awarded $100 Million to End Homelessness," MacArthur Foundation, April 7, 2021, https://www.macfound.org/press/article/community-solutions-awarded-$100-million-to-end-homelessness.

11. Brian Kenny, host, *Cold Call*, podcast, episode 174, "Can a Social Entrepreneur End Homelessness in the U.S.?," May 3, 2022, https://hbr.org/podcast/2022/05/can-a-social-entrepreneur-end-homelessness-in-the-u-s.

# Index